PIETRO NOLASCO

32 SHORTCUTS FOR 4 OPERATIONS

Arithmetic and algebra Strategies
with Vedic Mathematics

Quick math strategies
for middle and high school students
Exercises and solutions

ABOUT THE AUTHOR

Pietro Nolasco was born in Egypt on October 4, 1960, into an Italian family among the many Europeans who had settled in that country at that time. He works as a computer scientist for a public organization. He is passionate about mathematics and is a father who is attentive to the problems of his two children. He has never denied his help to his children, nephews, and children of friends in solving mathematical questions. Hence, he researched and developed fast calculation strategies that became winning methods to allow the rapid solution of calculations, even complicated ones.

After publishing the first book "Fast Calculation," he deepened the subject by undertaking a series of study courses at the prestigious Vedic Mathematics Academy (*www.vedicmaths.org*), where he obtained the "Advanced Diploma in Vedic Maths." He published an article in the "International Journal of Vedic Mathematics" of the Vedic Mathematics Academy. The journal collects articles by scholars and researchers of Vedic Mathematics from all over the world, which mostly concern experiences and discoveries of new computational methodologies.

He is a member of the IAVM (Institute for the Advancement of Vedic Mathematics, *https://instavm.org*), which aims to promote and disseminate Vedic Mathematics throughout the world.

He has published several popular brochures (www.calcoloveloce.it) to raise awareness about the potential and simplicity of Vedic Mathematics, highlighting its creativity, flexibility, and coherence. The approach is so effective that even those who are most apprehensive about mathematics come to love it.

Table of contents

32 SHORTCUTS FOR 4 OPERATIONS

INTRODUCTION

My primary school teacher taught me the first approach to calculation shortcuts. She taught me to multiply by 10, 100, 1000, and 11, thus avoiding the slowness of traditional calculation methods.

Over time, I have deepened my knowledge of the subject so much that today, calculating quickly and in my mind has become routine: I can get to the result sooner than it would take using a calculator.

Later, I discovered that this way of calculating is contemplated in the teachings of Vedic Mathematics.

Vedic mathematics differs from traditional mathematics in that the former uses rules that frequently require laborious and time-consuming procedures. Whereas the latter relies on mental calculations, effortlessly achieving faster results by following specific "paths" naturally established by the mind. These "paths" are condensed into sixteen aphorisms or "Sutras," each applicable to a broad spectrum of mathematical scenarios, all converging to the same mental "circuit." Consequently, numerous intricate calculations are executed swiftly, digit by digit, without stress, creating an almost magical perception of the method.

Calculation shortcuts come in handy every day. For example, they can be used to find out how much an item costs when a discount is applied, divide the cost of a dinner with friends, recalculate the ingredients of a recipe, and calculate on the fly how much change is due when you pay with a banknote.
For students, calculation shortcuts can help reach the result without getting lost in pages of steps or can even impress friends with the speed of computation. Complicated calculations can be completed easily on the blackboard, writing just two lines.
In this book, you will find the techniques explained directly with examples. Sometimes, mnemonic rules are mentioned to help remember how to perform the shortcut. These rules are none other than the aphorisms (or Sutras) on which Vedic mathematics is based.

At the end of the book, you will realize that a calculation can be performed in different ways, taking advantage of this or that shortcut. It will only be your mental predisposition that determines which method is most suitable for arriving at the result first.

I highly recommend performing the proposed exercises to internalize and automate the recognition of the type of calculation and the choice of relative shortcuts. You will find the solutions at the end of the book.

I hope this book will help you discover new, more fun ways to count and make the fear that many feel when faced with a calculation disappear, especially if it involves numbers with many digits.

Good luck with your math!

1. ADDITION - FIND ZERO AND SUM

Shortcut 1 –
The circle of 10...

The first shortcut to add more numbers is to group those whose sum is 10 or a multiple of 10. This way, you can have a number that ends with zero.

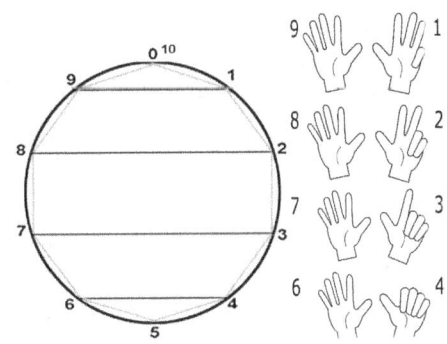

Circle of 10
The pairs of numbers that add up to 10 are:
10 + 0;   9 + 1;   8 + 2;   7 + 3;   6 + 4;   5 + 5

Example:

$$\underline{4} + \underline{7} + \underline{6} + 2 + \underline{3} = 10 + 10 + 2 = 22$$
$$|\underline{\quad 10 \quad}|$$
$$|\underline{\quad 10 \quad}|$$

3

A ten is the sum of 4 and 6.

Another ten is given by the sum of 7 and 3.

There remains 2 which, added to the two tens obtained, gives 22.

Example:
$$3\underline{7} + 1\underline{3} = 30 + \underline{7} + 10 + \underline{3} = 30 + 10 + 10 = 50$$

You immediately notice that units 7 and 3 are complementary (they add up to 10). Therefore, you add the tens and the sum of the units to make up another ten.

Example:
$$1\underline{4} + 2\underline{\underline{2}} + 27 + 1\underline{6} + 1\underline{\underline{8}} =$$
$$1\underline{4} + 1\underline{6} = 30$$
$$2\underline{\underline{2}} + 1\underline{\underline{8}} = 40$$
$$\text{Remains } 27$$
$$30 + 40 = 70; \quad 70 + 27 = 97$$

Example:

Look for the ten in the column:

(The numbers with the dot on the right identify the values that add up to 10)

```
  1             1            1  1            1  1
  2 4 .         3 5          4 . 6 . 2 .      3 . 1   5
+ 1 2         + 2 7 .      + 3 . 2   5 .    + 2 . 1   3
+ 3 3 .       + 1 2        + 7 . 3 . 3 .    + 7 . 8 . 9 .
+ 1 3 .       + 2 3 .      + 6 . 1 . 1      + 8 . 2 . 1 .
  8 2           9 7          2 1 3 1          2 1 3 8
```

Exercises

1.1 Add up looking for zero

a) $6 + 5 + 2 + 4 + 8 =$ $7 + 4 + 3 + 9 + 1 =$ $1 + 5 + 8 + 5 + 2 =$

b) $12 + 21 + 19 + 8 =$ $63 + 17 + 2 + 9 =$ $14 + 15 + 26 + 25 =$

c)

```
     2 1          4 4          1 5 3          2 1 6
   + 3 2        + 1 7        + 3 9 7        + 4 1 7
   + 1 8        + 1 3        + 7 5 4        + 4 8 4
   + 1 9        + 2 6        + 7 1 2        + 1 2 3
```

Shortcut 2 –
One less than the previous one, one more than the previous one

It applies to numbers close to those ending in zero.

Example:
$$89 + 7 = 90 + 6 = 96$$

In this case, 89 is close to 90.
If we add 1 to 89, we get 90, which ends with zero. Let's subtract 1 from 7: we are left with 6. We can now perform a simpler addition: $90 + 6 = 96$.

Example:
$$72 + 15 = 70 + 17 = 87$$

In this case, 72 is close to 70.
Let's subtract 2 from 72 to get 70. Adding 2 to 15, we obtain 17. We can now perform a simpler addition: $70 + 17 = 87$..

The rule is: 'The value that is subtracted or added to a number ending with zero is added or subtracted to the other number.'

Exercises

1.2 Search for zero by adding and subtracting values to numbers and summing them.

a) 29 + 41 =   32 + 23 =   18 + 22 =   33 + 38 =   99 + 6 =

b) 28 + 32 + 9 + 11 =   17 + 12 + 23 +18 =   89 + 91 + 8 + 12 =

c) 0.9 + 1.1 =   1.2 + 0.8 =   0.997 + 99.003 =   999.4 + 2.8

Shortcut 3 –
If the sum does not end with zero

What should you do if there are no numbers whose sum ends with zero?

If you obtain a **value that is 10 or greater** when adding two numbers, place a small circle beside the resulting digit and continue adding the re-maining units.

The benefit of this method is that it speeds up calculation by letting you utilize smaller numbers and eliminating the need to store carry.

Example:
Add in the column until you find or exceed ten:

$$
\begin{array}{r}
2\ 3\ 5\ 9 \\
+\ 5\ 5\ 8\ 7 \\
+\ 6\ 3\ 1\ 6 \\
+\ 1\ 3\ 8\ 2\ = \\
\end{array}
\Big)\ 9 + 7 = 16
$$

9 + 7 = 16

Insert a small circle next to 7. Consider only the **unit** 6 of 1<u>6</u> and add the next digit, 6:

6

$$
\begin{array}{r}
2\ 3\ 5\ 9 \\
+\ 5\ 5\ 8\ 7^\circ \\
+\ 6\ 3\ 1\ 6 \\
+\ 1\ 3\ 8\ 2\ =
\end{array}
$$
$\biggr)$ unit of 1<u>6</u>: 6
$6 + 6 = 12$

$6 + 6 = 12$

$$
\begin{array}{r}
2\ 3\ 5\ 9 \\
+\ 5\ 5\ 8\ 7^\circ \\
+\ 6\ 3\ 1\ 6^\circ \\
+\ 1\ 3\ 8\ 2\ =
\end{array}
$$
$\biggr)$ unità di 1<u>2</u>: 2
$2 + 2 = 4$

Insert a small circle next to 6. Consider only the unit 2 of 1<u>2</u> and add the next digit, 2:

$2 + 2 = 4$. Write 4:

$$
\begin{array}{r}
2\ 3\ 5\ 9 \\
+\ 5\ 5\ 8\ 7^\circ \\
+\ 6\ 3\ 1\ 6^\circ \\
+\ 1\ 3\ 8\ 2\ = \\
\hline
4
\end{array}
$$

The two circles in the first column represent the carry for the second column:

$\underline{2} + 5 = 7$; continue the addition until you exceed the ten;
$7 + 8 = 15$

Insert a small circle next to 8. Consider only the unit 5 of 15 and add the next digit, 1 and then 8:
$5 + 1 + 8 = 14$
Insert a small circle beside the second 8 and write 4.

$$
\begin{array}{r}
2\ 3\ 5\ 9 \\
+\ 5\ 5\ 8^\circ\ 7^\circ \\
+\ 6\ 3\ 1\ 6^\circ \\
+\ 1\ 3\ 8^\circ\ 2\ = \\
\hline
4\ \ 4
\end{array}
$$

The two small circles in the second column from the right represent the carry for the third column:
$\underline{2} + 3 + 5 = 10$. insert a small circle beside 5.
Consider the unit 0 of 10 and continue: $0 + 3 + 3 = 6$.

Write 6.

$$
\begin{array}{r}
2\ \ 3\ \ 5\ \ 9 \\
+\ 5\ \ 5^\circ\ 8^\circ\ 7^\circ \\
+\ 6\ \ 3\ \ 1\ \ 6^\circ \\
+\ \underline{1\ \ 3\ \ 8^\circ\ 2} = \\
6\ \ 4\ \ 4
\end{array}
$$

The small circle representing a carry for the left column is added to the remaining digits.:

$1 + 2 + 5 + 6 + 1 = 15$

$$
\begin{array}{r}
2\ \ 3\ \ 5\ \ 9 \\
+\ 5\ \ 5^\circ\ 8^\circ\ 7^\circ \\
+\ 6^\circ\ 3\ \ 1\ \ 6^\circ \\
+\ \underline{1\ \ 3\ \ 8^\circ\ 2} = \\
\mathbf{1\ \ 5\ \ 6\ \ 4\ \ 4}
\end{array}
$$

Exercises

1.3 Put a little circle for every ten that are found, then continue adding with the remaining units.

$$
\begin{array}{r}
3\ \ 2\ \ 7\ \ 4 \\
+\ 6\ \ 6\ \ 5\ \ 7 \\
+\ 2\ \ 3\ \ 1\ \ 4 \\
+\ \underline{3\ \ 3\ \ 2\ \ 1} =
\end{array}
\qquad
\begin{array}{r}
7\ \ 4\ \ 8\ \ 6 \\
+\ 1\ \ 5\ \ 4\ \ 1 \\
+\ 5\ \ 2\ \ 7\ \ 8 \\
+\ \underline{6\ \ 7\ \ 5\ \ 5} =
\end{array}
\qquad
\begin{array}{r}
3\ \ 4\ \ 3\ \ 7 \\
+\ 9\ \ 5\ \ 8\ \ 1 \\
+\ 2\ \ 7\ \ 5\ \ 4 \\
+\ \underline{6\ \ 2\ \ 3\ \ 8} =
\end{array}
$$

2. SUBTRACTIONS

Shortcut 4 –
All from nine, the last from ten.

Use this method when you want to subtract a number from a power of 10.

Example:

Subtract 346 from 1000:

$$1\,0\,0\,0$$
$$-\ \underline{3\,4\,6} =$$

You can perform calculations from left to right or right to left. The last digit will be the unit on the right-hand side.
Applying the rule: *All from nine, the last from ten*, you will have:

$$
\begin{array}{ccc}
9- & 9- & 10- \\
\underline{3} & \underline{4} & \underline{6} \\
\mathbf{6} & \mathbf{5} & \mathbf{4} \\
|\rule{1.5cm}{0.4pt}| & |\rule{0.5cm}{0.4pt}| & \\
\text{subtract} & \text{\textit{subtract}} & \\
\text{from 9} & \text{\textit{from 10}} &
\end{array}
$$

Two numbers are complementary to 10 if their sum is 10.
For example, 6 and 4 are complementary to 10 because their sum is 10.
Similarly, two numbers are complementary to 9 if their sum is 9.

Example: 5 and 4 are complementary to 9 because btheir sum is 9.

With the exception of the final digit, which calculated the 10's complement, all of the digits on the left in this example have their 9's complements determined.

Similary, the 1000's complement of 346 is 654.
(In fact 346 + 654 = 1000).

When you have to figure out how much change you'll have after buying, this method comes in handy.

Example:

Suppose you wish to use a 10-euro banknote to pay for an item that costs 8.76 euros. In this instance, it is also necessary to provide the decimal figures (i.e., cents).

$$
\begin{array}{r}
1\,0.\,0\,0 \\
-\ \underline{8.\,7\,6} =
\end{array}
$$

Calculating from left to right we have:

$$9 - 8 = \mathbf{1} \qquad 9 - 7 = \mathbf{2} \qquad 10 - 6 = \mathbf{4}$$

$$
\begin{array}{r}
1\,0.\,0\,0 \\
-\ \underline{8.\,7\,6} = \\
\mathbf{1.\,2\,4}
\end{array}
$$

A second real-world scenario involves two percentage-expressed values, of which one is known. Here, we have to figure out the other one.

Example:

Assume that girls attend a school in proportion of 62%. What proportion of boys are there?

We need to calculate the complement to 100 of 62.

$$9 - 6 = \mathbf{3} \qquad 10 - 2 = \mathbf{8}$$

Therefore, the percentage of boys in the school is 38%.

Example:

Determine the complement to 1000 of 89.
We write 89 with three digits by moving the zero to the left, or 089, since 1000 has three zeros.
The complement to 1000 of 089 is:
$$9 - 0 = 9; \ 9 - 8 = 1; \ 10 - 9 = 1$$
We get: **911**.

Example:

Determine the complement to 1000 of 450.
Calculate the complement of 45 without taking into consideration the zero at the end of 450. Finally, add the zero to the right-hand side.
We can deduct 4 from 9 and 5 from 10 to get the complement to 100 of 45. After that, we put 0 to the left-hand side:

$$9 - 4 = 5 \qquad 10 - 5 = 5$$
So, we obtain: **550**

Exercises

2.1 Using the proper complement, determine the following differences:

a) $10 - 6 =$ $100 - 27 =$ $100 - 74 =$ $100 - 43 =$ $100 - 25 =$

b) $1000 - 238 =$ $1000 - 127 =$ $1000 - 790 =$ $1000 - 77 =$

c) $10000 - 2456 =$ $10000 - 7942 =$ $10000 - 322 =$

d) $100{,}00 - 24{,}38 =$ $1000{,}00 - 326{,}84 =$ $1000{,}00 - 52{,}09 =$

Shortcut 5 –
Subtract by adding

This technique works well when the amount to be subtracted is almost a **power of 10**.

Example:
$$186 - 97$$
The subtrahend 97 is close to 100. To reach 100, we need to add 3.
(The complement to 100 of 97 is 3).
Adding 3 to 97, you get 100. In order to prevent an incorrect subtraction result, you also need to add 3 to the minuend 186. So we obtain an easy sutraction: 189 – 10 = 89.

$$186 + 3 - \qquad 189 -$$
$$\rightarrow$$
$$\underline{97 + 3} = \qquad \underline{100} =$$
$$89$$

Quickly: 186 minus 100 + complement of 97, which is 3:
$$186 - 100 + 3 = 89$$

Example:
$$1657 - 839 =$$
The subtrahend 839 is almost at 1000; 161 more is needed to get there.
(To obtain the complement to 1000 of 839, we can use the rule 'All from nine, the last from 10':
$$9 - 8 = 1 \quad 9 - 3 = 6 \quad 10 - 9 = 1).$$
Subtraction becomes:

$$1657 + 161 \qquad 1818$$
$$\rightarrow$$
$$- \underline{839 + 161} = \qquad - \underline{1000} =$$
$$818$$

Quickly: 1657 minus 1000 + complement of 839, which is 161:
$$1657 - 1000 + 161 =$$
$$657 + 161 = 818$$

Example:
$$2887 - 1642 =$$
We take off 1000 from both terms as they both include 1000, and then we add the complement to 1000 of the remaining subtrahend.
Subtraction becomes:

$$2887 - 1000 \qquad 1887 -$$
$$\rightarrow$$
$$- \underline{\ 1642 - 1000} = \quad - \underline{\ 642} =$$

2887 minus twice 1000 + complement to 1000 of 642, which is 358:

The complement to 1000 of 642 is 358

$$1887 + 358 \qquad 2245$$
$$\rightarrow$$
$$- \underline{\ 642 + 358} = \quad - \underline{1000} =$$
$$\qquad\qquad\qquad\qquad 1245$$

Quickly:
$$2887 - 2000 + 358 =$$
$$887 + 358 = 1245$$

Example:
$$4978 - 3642 =$$
Both numbers contain 3000, so we can perform the subtraction quickly:
$$4978 - 3000 - 1000 + \textit{complement to } 1000 \textit{ of } 642, \textit{ which is } 358:$$
$$4978 - 4000 + 358 =$$
$$978 + 358 = 1336$$

Example:
$$6742 - 4527 =$$
Quickly: *Both numbers contain 4000 so:*

$$6742 - 4000 - 1000 + \textit{complement to } 1000 \textit{ of } 527, \textit{ which is } 473:$$
$$6742 - 5000 + 473 =$$
$$1742 + 473 = 2215$$

Exercises

2.2 Solve the following subtractions using the sum method:

a) $156 - 98 = \qquad 237 - 96 = \qquad 318 - 93 = \qquad 579 - 389 =$

b) $1537 - 994 = \qquad 2729 - 1971 = \qquad 5494 - 3877 = \qquad 4568 - 2877 =$

Shortcut 6 –
Subtract without the carry
using the complement of 10

This is a quick method that makes use of two guidelines:
- *All from nine, last from 10 (10's complement).*
- *One less than the previous one.*

Example:

$$6\ 7\ 4\ 2$$
$$-\ \ \underline{4\ 9\ 2\ 8}\ =$$

Making a right-to-left calculation:
There is no positive natural number obtained from $2 - 8$. You can compute $8 - 2 = 6$, then determine the complement to 10 of 6, which is 4.

Write 4.

$$6\ 7\ 4\ 2$$
$$-\ \ \underline{4\ 9\ 2\ 8}\ =$$
$$4$$

$8 - 2 = 6$
$10 - 6 = 4$

Because the complement technique was applied, you must subtract one from the previous digit of the minuend. After performing $4 - 1 = 3$, you can proceed to the next subtraction: $3 - 2 = 1$. Write 1.

$$\overset{3}{6}\ 7\ 4\ 2$$
$$-\ \ \underline{4\ 9\ 2\ 8}\ =$$
$$1\ 4$$

$4 - 1 = 3$
$3 - 2 = 1$

$7 - 9$ does not result in a positive natural number. .
Calculate $9 - 7 = 2$ and then the complement to 10 of 2, which is 8.
Write 8.

$9 - 7 = 2$
$10 - 2 = 8$

$$\overset{3}{6}\ 7\ 4\ 2$$
$$-\ \ \underline{4\ 9\ 2\ 8}\ =$$
$$8\ 1\ 4$$

Because the complement technique was applied, you must subtract one from the previous digit of the minuend. After performing $6 - 1 = 5$, you can proceed to the next subtraction: $5 - 4 = 1$. Write 1.

$$
\begin{array}{r}
\overset{5}{6}\,\overset{3}{7}\,4\,2 \\
-\ 4\,9\,2\,8 = \\
\hline
1\,8\,1\,4
\end{array}
$$

| 6 − 1 = 5 |
| 5 − 4 = 1 |

Exercises

2.3 Practice the following subtractions:

a)
$$
\begin{array}{r} 6\ \ 3 \\ -\ 3\ \ 7 = \\ \hline \end{array}
\qquad
\begin{array}{r} 8\ \ 2 \\ -\ 6\ \ 6 = \\ \hline \end{array}
\qquad
\begin{array}{r} 7\ \ 9 \\ -\ 4\ \ 1 = \\ \hline \end{array}
\qquad
\begin{array}{r} 9\ \ 9 \\ -7\ \ 2 = \\ \hline \end{array}
$$

b)
$$
\begin{array}{r} 6\ \ 2\ \ 4 \\ -\ 3\ \ 3\ \ 5 = \\ \hline \end{array}
\qquad
\begin{array}{r} 8\ \ 5\ \ 7 \\ -\ 3\ \ 7\ \ 9 = \\ \hline \end{array}
\qquad
\begin{array}{r} 6\ \ 9\ \ 1 \\ -\ 3\ \ 3\ \ 7 = \\ \hline \end{array}
$$

c)
$$
\begin{array}{r} 5\ \ 3\ \ 3\ \ 4 \\ -\ 3\ \ 4\ \ 5\ \ 6 = \\ \hline \end{array}
\qquad
\begin{array}{r} 6\ \ 4\ \ 6\ \ 7 \\ -\ 3\ \ 8\ \ 3\ \ 5 = \\ \hline \end{array}
\qquad
\begin{array}{r} 6\ \ 7\ \ 4\ \ 4 \\ -\ 5\ \ 2\ \ 8\ \ 2 = \\ \hline \end{array}
$$

Shortcut 7 –
Subtract without the carry by removing and adding 1

This is a quick method that takes advantage of two rules:
- *One less than the previous.*
- *One more than the previous one.*

Example:

Let's have the following subtraction:

$$
\begin{array}{r} 3\ \ 7\ \ 9 \\ -2\ \ 8\ \ 7 = \\ \hline \end{array}
$$

The first thing to do is to look at the first digit on the top left: 3 (of 379).
The digit 3 represents the hundreds in 379, so 3 hundred are 300 units.

Subtract one from this value: $300 - 1 = 299$

Perform the subtraction between the found number 299 and the subtrahend 287:

> From 379 take 300
> and subtract 1
> from it:
> $300 - 1 = 299$

(minuend 379)

$$\begin{array}{r} 2\ 9\ 9 \\ -\ 2\ 8\ 7\ = \\ \hline 0\ 1\ 2 \end{array}$$

> From 299 subtract 287
> obtaining 012

Consider the minuend stripped of the first digit on the left and increased by 1:

> From 379 delete the 3:
> 79 remains

$$3\,7\,9 \rightarrow 79$$
$$79 + 1 = 80$$

> Add 1 to 79
> obtaining 80

Add 80 to 012:

$$\begin{array}{r} 2\ 9\ 9 \\ -\ 2\ 8\ 7 \\ \hline 0\ 1\ 2\ + \\ 8\ 0\ = \\ \hline 9\ 2 \end{array}$$

> Add 80 to 012
> obtaining the result 92

The trick works because first 79 was subtracted from 379 to get 300 and then 1 was subtracted to get 299. Finally, $79 + 1$ was performed to reba-lance the operation.

Summarizing:

You can calculate the value of the first digit on the left-hand side as a power of 10 and then subtract 1 from its value.

After that, you can perform the subtraction using this result.

Finally, you can add the value you get from the first term of the subtraction without the first digit on the left, increasing it by 1.

Example:

$$4\ 5\ 6\ 7$$
$$-\ 1\ 8\ 3\ 4\ =$$

The first digit to the left of the first row is 4. The corresponding power of 10 is 4000 because 4 represents the thousands of the minuend.

$4000 - 1 = 3999$

We get:

(minuend 4567)

$$3\ 9\ 9\ 9$$
$$-\ 1\ 8\ 3\ 4\ =$$

$$2\ 1\ 6\ 5$$

4̸ 5 6 7

Perform $567 + 1 = 568$:

$$2\ 1\ 6\ 5$$
$$+\ \ \ 5\ 6\ 8\ =$$

$$2\ 7\ 3\ 3$$

Observe how this method eliminates the need for carries by applying subtraction from 9 to all digits except the first.

Exercises

2.4 Work out with the following subtractions:

a)

$$\begin{array}{c}7\ 2\ 3\ 6 \\ -\ 3\ 5\ 8\ 7\ = \end{array} \qquad \begin{array}{c}4\ 6\ 5\ 7 \\ -\ 2\ 4\ 3\ 3\ = \end{array} \qquad \begin{array}{c}6\ 9\ 3\ 4 \\ -\ 5\ 4\ 6\ 8\ = \end{array}$$

b)

$$\begin{array}{c}5\ 4\ 5\ 8 \\ -\ 2\ 6\ 7\ 6\ = \end{array} \qquad \begin{array}{c}7\ 6\ 4\ 9 \\ -\ 2\ 3\ 5\ 5\ = \end{array} \qquad \begin{array}{c}8\ 7\ 6\ 5\ - \\ -\ 5\ 3\ 7\ 9\ = \end{array}$$

3. MULTIPLYING AND DIVIDING BY 9, 11, 111

Shortcut 8 –
Multiplying by 9

This technique works well for numbers with many digits.
It involves simple subtractions without any multiplications.
The rule *All from nine, the last from ten* can also be applied.

Example:

$$679 \times 9 =$$

Rewrite the number with a zero preceding it and one following it:

$$0\ 6\ 7\ 9\ 0$$

(With practice, the added zeros will only show up in your mind).

Starting with the second digit from the left (6), subtract the one before it:

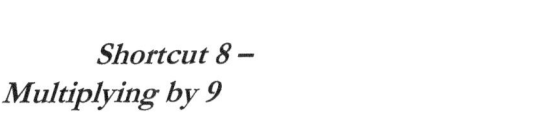

$$0\quad 6\quad 7\quad 9\quad 0$$
$$6 - 0 = 6$$

You can continue in the same way by scaling one digit to the right:
7 - 6 = 1. Write 1 after the 6:

$$6\ 1$$

Proceed by scaling rightward: 9 - 7 = 2. Write 2:

$$6\ 1\ 2$$

Last subtraction: $0 - 9 = (-9)$.

(It turns out to be negative. Calculate 9 - 0 and place minus sign in front).

$$\mathbf{6\ 1\ 2\ (-\ 9)}$$

The negative number might be written with an overline for ease of reading:

(Such a number is called a "bar number" or "vinculum").

$$\mathbf{6\ 1\ 2\ \overline{9}}$$

To find the result subtract 9:

$$6120 - 9 = \mathbf{6111}$$

Quickly: *Calculate the complement to ten of the overline figure ($\overline{9}$) and decrease the preceding digit (2) by 1 :*

The complement to 10 of 9 is **1**; Preceding figure decreased by one: $2 - 1 = \mathbf{1}$

$$612\overline{9} = \mathbf{6111}$$

Example:

$$7649 \times 9 =$$

$$\leftarrow \quad \leftarrow \quad \leftarrow \quad \leftarrow \quad \leftarrow$$

$$\mathbf{\textit{0}\quad 7 \quad 6 \quad 4 \quad 9 \quad \textit{0}}$$

(When the subtraction is negative put an overline over the digit)

$7 - 0 = \mathbf{7}$;	**6 – 7 is negative.**
$6 - 7 = \mathbf{\overline{1}}$;	**Calculate 7 – 6 and bar**
$4 - 6 = \mathbf{\overline{2}}$;	**out the result: $\overline{1}$**
$9 - 4 = \mathbf{5}$;	**Similarly**
$0 - 9 = \mathbf{\overline{9}}$	$4 - 6 = \overline{2}$
	$0 - 9 = \overline{9}$

$$7\overline{12}5\overline{9}$$

The number can be broken down into two groups: $7\overline{12} \mid 5\overline{9}$

To find the result, subtract from the left: $700 - 12 = 688$ and $50 - 9 = 41$

We get: **68841**

Quickly: *Write the complement of the overlined digits and decrease the preceding figure by 1 :*

The complement to 100 of 12 is 88 *(All from nine, the last from ten); the previous figure decreased by 1 is $7 - 1 = 6$. The complement to 10 of 9 is 1; the previous figure decreased by 1 is $5 - 1 = 4$. Result: $7\overline{12}5\overline{9} = 68841$.*

Note: *Since the last subtraction is always negative (except when it is zero), decrease the penultimate result by one and write the complement to ten of the last digit.*

Esercizi

3.1 Multiply by 9

a) 458 x 9 =   2578 x 9 =   35471 x 9 =   726846 x 9 =

b) 5489 x 9 =   76590 x 9 =   974963 x 9 =   6786543 x 9 =

Shortcut 9 –
Multiplying by 11 and by 111

Multiplication by 11

One method of multiplying by 11 is to multiply the number by ten and add the number itself.

Example:

$$23425 \times 11 =$$

$$23425 \times 10 + 23425$$

Multiplying by ten is equivalent to adding a zero to the right of the number. We have:

$$
\begin{array}{r}
234250 \\
+ \ \underline{23425} \\
257675
\end{array} =
$$

You can see that the result has as its first and last digits the corresponding first and last digits of the factor to be multiplied by 11 (*23425*).

Intermediate figures are given by the sum of each term with the next:

```
    +  + + +
  2  3 4 2 5
  / \/ \/ \/ \/ \
  2  5 7 6 7 5
```

2;   2 + 3 = 5;   3 + 4 = 7;   4 + 2 = 6;   2 + 5 = 7;   5

257675

Thus, for multiplications by 11, we have a particular computation technique..

Example;

$$34162 \times 11 =$$

+ + + +
3 4 1 6 2
/ \/ \/ \/ \/ \
3 7 5 7 8 2

> 1st digit on left 3
> 3 + 4 = 7
> 4 + 1 = 5
> 1 + 6 = 7
> 6 + 2 = 8
> Last digit   2

The next example shows how the carry is handled:

Example:

$$34572 \times 11 =$$

+ + + +
3 4 5 7 2
/ \/ \/ \/ \/ \
3 7 9 ^{1}2 9 2
3 8 0 2 9 2

> 1st digit on left 3
> 3 + 4 = 7
> 4 + 5 = 9
> 5 + 7 = 12 carry 1
> 7 + 2 = 9
> Last digit   2

(In this example, one of the sums results 5 + 7 = 12. The unit (2) is written and the ten (1) is the carry over to the previous digit 9.
9 plus the carry gives 10. 0 is written, and 1 is carried over to the previous digit, 7.
7, plus the carry, becomes 8).

Moltiplication by 111

Example:

$$\begin{array}{r} 3\,2\,1 \\ \times \ \underline{1\,1\,1} = \end{array}$$

Write the first digit on the right-hand side: 1

3 2 1
        \
          1

Add up the first two digits on the right: 2 + 1 = 3

22

$$\begin{array}{ccc} 3 & 2 & 1 \\ & \boxed{+} & \\ & 3 & 1 \end{array}$$

Sum the three digits: $3 + 2 + 1 = 6$

$$\begin{array}{ccc} 3 & 2 & 1 \\ \boxed{+} & \boxed{+} & \\ 6 & 3 & 1 \end{array}$$

Add the two digits on the left: $3 + 2 = 5$

$$\begin{array}{ccc} 3 & 2 & 1 \\ \boxed{+} & & \\ 5 & 6 & 3 & 1 \end{array}$$

Write the first digit on the left-hand side: **3**

$$\begin{array}{ccc} 3 & 2 & 1 \\ / & & \\ 3 & 5 & 6 & 3 & 1 \end{array}$$

35631

Example with then carry:

$$4\ 8\ 5\ \times\ 1\ 1\ 1\ =$$

Last digit: **5**; $8 + 5 = 13$, write **3**, carry 1:

$$\begin{array}{ccc} 4 & 8 & 5 \\ & & \diagdown \\ & _1 3 & 5 \end{array}$$

Add the three digits plus the carry: $4 + 8 + 5 + 1 = 18$;
write **8**, carry 1:

$$\begin{array}{ccc} 4 & 8 & 5 \\ _1 8 & _1 3 & 5 \end{array}$$

Add the two digits on the left plus the carry: $4 + 8 + 1 = 13$; write **3**,
carry 1:

$$\begin{array}{ccc} 4 & 8 & 5 \\ _1 3 & _1 8 & _1 3 & 5 \end{array}$$

Write the first digit on the left added to the carry: $4 + 1 = $ **5.**

$$\begin{array}{ccc} 4 & 8 & 5 \\ / & & \\ 5 & _1 3 & _1 8 & _1 3 & 5 \end{array}$$

53835

(*Big numbers can also follow this rule*).

Exercises

3.2 Multiply

a) 245 x 11 =   2536 x 11 =   35461 x 11 =   736725 x 11 =

b) 6574 x 11 =   82670 x 11 =   944785 x 11 =   6875363 x 11 =

c) 622 x 111 =   826 x 111 =   756 x 111 =   687 x 111 =

Shortcut 10 –
Dividing by 9

When we divide a number by nine, for each ten of the dividend, we obtain a nine and a remainder of one.

The units of the dividend contribute to the remainder.

Example:

$$32 \div 9 = 3 \quad \text{remainder} = 5$$

There are three tens, each containing a nine and a one as a remainder, so we get three nines and three ones.

The three ones have to be added to the units (2 of **32**) to get the final remainder.

$(3 + 2 = 5)$.

Consequently, we have a pattern of calculation: we find the result by writing down the first digit to the left of the dividend. Adding it to the units, we get the remainder.

$$3\ 2 \div 9 =$$

(3+2)

3 | 5 (remainder)

1st digit on the left 3
3 + 2 = 5
Remainder = 5

Example:

$$5\ 7 \div 9 =$$

(5 + 7)

5 | 12 remainder

1st digit on the left 5
5 + 7 = 12
Remainder = 12

Since 12 is greater than 9, subtract 9 from the remainder and increase the quotient by 1:

$12 - 9 = 3$ *and* $5 + 1 = 6$

$$57 \div 9 = 6 \quad \text{remainder} = 3$$
$$\text{or even } 6.33333\ldots$$

Note: *If there is a remainder when dividing by 9, the outcome is the quotient plus a decimal portion with an infinite number of digits corresponding to the remainder (we get a recurring decimal number with repeating decimals corresponding to the digit of the remainder).*

The pattern remains valid even with dividends with many digits:
    1 - write in the quotient line the first digit to the left,
    2 - add this to the next figure of the dividend,
    3 - if the value obtained is greater than or equal to 9, subtract 9 (or more than one 9) and add the carry to the previous digit of the result.
    4 - repeat the cycle until you run out of digits. The last value is the remainder.

Example:

$$3\,2\,4\,7 \div 9 =$$

Report the first digit on the left:

3

> 3 2 4 7
> 1ˢᵗ digit on left 3
> 3

Sum this value by the next digit: $3 + 2 = \mathbf{5}$

$$3\,2\,4\,7 \div 9 =$$

3 5

> 3 2 4 7
> 3
> $3 + 2 = 5$

Sum this value by the next digit: $5 + 4 = 9$.
Since the result is more than 8, subtract 9 and add 1 to the preceding digit.

> 3 2 4 7
> 3 5
> $5 + 4 = 9$
> $9 - 9 = 0$ *carry 1*

$9 - 9 = 0$ carry *1*

$$3\,2\,4\,7 \div 9 =$$

> 3 2 4 7
> 3 5¹ 0
> $0 + 7 = 7$

1
3 5 0

The last operation provides the remainder.

Remainder: $0 + 7 = 7$

$$3247 \div 9$$

1
3 5 0 | 7

Add the carry to 5:

3 6 0 Resto = 7

o anche 360.7777...

Example:

$$6538 \div 9 =$$

Firs digit: 6;

6

$6 + 5 = 11$ (greather than 8). $11 - 9 = 2$ the carry is 1;

$$6538 \div 9$$

1
6 2

$2 + 3 = 5$ and $5 + 8 = 13$ (greather than 8). $13 - 9 = 4$ the carry is 1:

$$6538 \div 9$$

1   1
6 25 | 4

Add the carries:

7 2 6 Remainder = 4

or   726.444...

Note: By adding up the digits of the dividend several times until you get one figure, you get the remainder of the division. If this is zero or nine, the dividend is divisible by 9. In the above example we have: $6 + 5 + 3 + 8 = 22$ and $2 + 2 = 4$ which is the remainder.

Exsercises

3.3 Divide by 9

$55 \div 9 =$

$122 \div 9 =$

$3123 \div 9 =$

$4253 \div 9 =$

$28363 \div 9 =$

Shortcut 11 –
Dividing by *11*

Example:

$386 \div 11$

Write, on the second line, the first digit to the left of the dividend:

$$3\ 8\ 6 \div 11$$

$$\downarrow$$

$$3$$

Subtract from the next digit of the dividend the value written on the second line. Write the result on the second line to the right of the first value: $(8 - 3 = 5)$.

$$3\ 8\ 6 \div 11$$

$$\swarrow\ -$$

$$3\ 5$$

Repeat this procedure until you have considered all the digits of the dividend. The last number written is the remainder of the division: *(6 – 5 = 1)*.

$$3\ 8\ 6 \div 11$$

$$\downarrow\!-\!\searrow$$

$$3\ 5\ \ R\ 1$$

$$386 \div 11 = 35 \quad \textbf{Reminder} = 1$$

The result of dividing a number by eleven has a two digits recurring decimal number.
*To find its value, **multiply the remainder by nine.***

386 ÷ 11 = 35 remainder =**1**; Recurring decimals: 1 x 9 = **09**
We can also write:

$$386 ÷ 11 = 35.0909...$$

Few more rules to follow:

IThere are situations where there could be a negative difference between the dividend digit and the value on the second line. You can write the negative number either overlined (bar number) or in parentheses.

*When the **remainder is positive**, if it is worth **11**, you take it to **zero** and **increase the quotient by one.***
*If it is **more than 11**, you take **11** away from the remainder and increase the quotient by one.*
*In the event that the **remainder is negative**, you will **subtract one from the previous digit** and **compute the complement to 11**.*

Example:

$$957 ÷ 11$$

Write the first digit on the left:

9 5 7
9

$5 - 9 = -4$ which is reported as a bar number $\overline{4}$:

9 5 7
9 $\overline{4}$

$7 - (-4) = 7 + 4 = 11$.

9 5 7
9 $\overline{4}$ 11

Since 11 is the remainder of a division by 11, put it to zero and increase the quotient by one:

9 $\overline{4}$, 0

Add $\overline{4}$ and 1: $\overline{4} + 1 = \overline{3}$

9 $\overline{3}$ 0

Transform it to a positive number. You get:

28

$$90 - 3 = 87$$

87   Remainder = 0

*(**Quickly**: Write the complement to 10 of 3, which is 7, and decrease the preceding digit by one: 9 becomes 8, $\overline{3}$ becomes 7).*

Example:

$$8756 \div 11$$

Write the first digit (8) to the left of the dividend on the second line.

Subtract from the second digit of the dividend (7) the value on the second line (8):

(In this example, as in the following examples, negative figures are overlined).

$$7 - 8 = -1 = \overline{1}$$

$$8 \quad 7 \quad 5 \quad 6 \div 11$$
$$8 \quad \overline{1}$$

Calculate $5 - (-1) = 5 + 1 = 6$;   following: $6 - 6 = 0$.

$$8 \quad 7 \quad 5 \quad 6 \quad \div 11$$
$$8 \quad \overline{1} \quad 6 \quad 0$$
$$8\overline{1}6 \quad \text{Remainder} = 0$$

Transfor the result to an all-positive number:

$$8\overline{1}\,|\,6$$
$$80 - 1\,|\,6$$

796   Remainder = 0

*(**Quickly**: The complement to 10 of 1 is worth 9. Decrease the preceding digit by one).*

Example:

$$98765 \div 11$$

The first digit on the left is reported (9).

$$9 \ 8 \ 7 \ 6 \ 5 \div 11 =$$

$$9$$

Proceed with the subsequent subtractions:

$$8 - 9 = -1 = \overline{1}; \qquad 7 - (-1) = 7 + 1 = 8;$$
$$6 - 8 = -2 = \overline{2}; \qquad 5 - (-2) = 5 + 2 = 7$$
$$9\ 8\ 7\ 6\ 5 \div 11 =$$
$$9\ \overline{1}\ 8\ \overline{2}\ \ r : 7$$
$$9\overline{1}8\overline{2} \quad Remainder = 7$$
$$9\overline{1}\,|\,8\overline{2} =$$
$$90 - 1\,|\,80 - 2 \quad or$$

Quickly: *The complement to 10 of 1 is worth 9. Decrease the preceding digit by one: 9 − 1 = 8. The complement to 10 of 2 is worth 8. Decrease the preceding digit by one: 8 − 1 = 7.*

8978   Remainder = 7

in decimal number: *(7 **X** 9 = 63):* 8978.6363...

Example:

$$827421 \div 11$$

First digit: *8;* $\quad 2 - 8 = \overline{6};$ $\quad 7 - (-6) = 13;$ $\quad 4 - 13 = \overline{9};$
$2 - (-9) = 11;$ $\quad 1 - 11 = \overline{10}$

When you have two-digit results, you put the carry on the digit that precedes:

$$8 \quad 2 \quad 7 \quad 4 \quad 2 \quad 1$$
$$8 \quad \underline{\overline{6}}_{\,1}3 \quad \underline{\overline{9}}_{\,1}1 \quad \overline{10}$$

Add the remainders appropriately:

$$-6 + 1 = 5 \quad e \quad -9 + 1 = -8 \quad \text{we get:}$$
$$8\overline{5}3\overline{8}1 \quad Remainder = \overline{10}$$

Convert to a positive number: $8\overline{5}\,|\,3\overline{8}\,|\,1 = \ 80 - 5\,|\,30 - 8\,|\,1$

$$75221 \quad Remainder = \overline{10}$$

The **remainder is negative**, so calculate the **complement to 11** for the remainder and **decrease the quotient by one**:

$$75221 - 1 = 75220 \qquad \text{Remainder} = 11 - 10 = 1$$

$$75220 \text{ Remainder} = 1 \text{ or}$$

$$\textbf{75220.0909...}$$

You can also calculate it: 8;
2 – 8 = $\overline{6}$;
7 – (– 6) = 13 where 13 – 11 = 2 reminder = 1;
4 – 2 = 2;
2 – 2 = 0;
1 – 0 = 1
Pattern: 8 2 7 4 2 1 ÷ 11 = 8 $\overline{6}$, 2 2 0   r = 1   → 75220.0909...

Note. *A number is divisible by 11 if the difference between the sum of the odd-order digits and the even-order digits (counted from right to left) is 0, 11, or a multiple of it.*

The number 1 2 3 4 3 2 1 is divisible by 11 because the sum of the odd-order digits is (counting from right to left) 1+3+3+1= 8.
Still counting from right to left   1 2 3 4 3 2 1, the sum of the even-order digits is 2+4+2 = 8.
The difference between the two results is 8 – 8 = 0.

Exercises

3.4 Divide by 11

a) $87 \div 11 =$ $319 \div 11 =$ $987 \div 11 =$ $785 \div 11 =$

b) $2.645 \div 11 =$ $7367 \div 11 =$ $9896 \div 11 =$ $14523 \div 11 =$

c) $78412 \div 11 =$ $56734 \div 11 =$ $92779 \div 11 =$ $968367 \div 11 =$

Shortcut 12 –
Divide by 11 on the row

Example:

$$87 \div 11$$

Performing the division in mind, we have that $87 \div 11 = \mathbf{7}$ with **remainder = 10**. Representing the division as a fraction, we get:

$$\frac{87}{11} = 7 + \frac{10}{11}$$

To calculate the decimal digits of $\frac{10}{11}$ we multiply the numerator (10) by nine.

The values obtained will have two digits and will be placed as the **recurring decimal part**:

$$10 \times 9 = 90$$

$$\frac{87}{11} = 7,\dot{9}\dot{0}$$

Example:

$$\mathbf{234 \div 11}$$

11 into 23 is contained twice with one remaining;

Write the remainder (1) before the next digit of the dividend (4). 14 is formed.

$$23_14 \div 11 = 2....$$

11 into 14 is contained once, with the remainder of 3:

$$23_14 \div 11 = 21 \quad \text{Resto } 3$$

Calculate the decimal part: since $3 \times 9 = 27$, we have:

$$\frac{234}{11} = 21_1\frac{3}{11} = 21,\dot{2}\dot{7}$$

Example:

$$3376 \div 11$$

11 into 33 is contained three times, leaving a remainder of 0;

$$33_076 \div 11 = 3.....$$

11 into 07 is contained 0 times with a remainder of 7;

$$33_07_76 \div 11 = 30.....$$

11 into 76 is contained 6 times with a remainder of 10;

$$3_07_76 \div 11 = 306 \quad \text{Remainder } 10$$

Since 10 **x** 9 = 90, we have

$$\frac{3376}{11} = 306 + \frac{10}{11} = 306,\dot{9}\dot{0}$$

Exercises

3.5 Divide:

a) $96 \div 11 =$ $218 \div 11 =$ $786 \div 11 =$ $985 \div 11 =$

b) $2475 \div 11 =$ $7562 \div 11 =$ $8837 \div 11 =$ $15573 \div 11 =$

c) $68524 \div 11 =$ $51754 \div 11 =$ $82679 \div 11 =$ $987357 \div 11 =$

About divisions by 111, refer to Shortcut 32: Divisions close to a Base.

4. DOUBLE AND HALVE

Doubling a number means adding it to itself, that is, multiplying it by two. So it is an easy operation that you can do mentally.
However, people may encounter some difficulty when the values are a l are slightly elevated.
The following shortcut approaches the mental calculation.
You compute the doubling shortly, focusing on the most significant digits, which, in practice, involves moving from left to right.

Shortcut 13 –
Multiplying by 2 from left to right

(Useful for mental computation)

When the digits of the number to be doubled are **small**, that is, they are worth **1, 2, 3, or 4,** the calculation proceeds expeditiously without a carry. On the other hand, when you have a number to double which contains **big digits**, i.e., they are worth **5, 6, 7, 8, or 9**, these generate the carry of a ten.

Example:
Multiplication by 2 using **small digits**:

$$
\begin{array}{r}
4\,2\,3\,1\,\times \\
2\,= \\
\hline
8\,4\,6\,2
\end{array}
$$

You can double digit by digit without further calculations, so the outcome is instantaneous.

Example:
Multiplication by 2 with **big digits:**

$$
\begin{array}{r}
3\,7 \\
\times \underline{\quad 2} =
\end{array}
$$

From left:
Multiply 3 x 2 = 6. But, **before writing** the result, **"peek" at the next number** 7 **to the right:**
If it is a **small figure**, write down the result so far.
If it is a **big digit, add 1** and write the result.
Since 7 is a big digit, calculate 6 + 1 = 7 and write the result:

$$
\begin{array}{r}
3\,7 \\
\times \underline{\quad 2} = \\
7
\end{array}
$$

Multiply 7 x 2 = 14 and write the unit $\underline{4}$ of 1$\underline{4}$:

$$
\begin{array}{r}
3\,7 \\
\times \underline{\quad 2} = \\
7\,4
\end{array}
$$

Example:

$$
\begin{array}{r}
4\,9\,3\,7 \\
\times \underline{\qquad 2} =
\end{array}
$$

Multiply 4 x 2 = 8. Since the next digit 9 is big, calculate
8 + 1 = 9. Write 9:

$$
\begin{array}{r}
4\,9\,3\,7 \\
\times \underline{\qquad 2} = \\
9
\end{array}
$$

Multiply 9 x 2 = 18. Since the next digit 3 is small, write the unit $\underline{8}$ of 1$\underline{8}$:

$$
\begin{array}{r}
4\,9\,3\,7 \\
\times \underline{\qquad 2} = \\
9\,8
\end{array}
$$

Multiply 3 x 2 = 6. Since the next digit 7 is big, calculate
6 + 1 = 7. Write 7:

$$
\begin{array}{r}
4\,9\,3\,7 \\
\times \underline{\qquad 2} = \\
9\,8\,7
\end{array}
$$

Multiply 7 ✗ 2 = 14. Since it is the last digit, write the unit 4 of 14:

$$
\begin{array}{r}
4\ 9\ 3\ 7 \\
\times \underline{\qquad 2} = \\
9\ 8\ 7\ 4
\end{array}
$$

Example:

Multiplication by 2 with **big digits**:

$$
\begin{array}{r}
8\ 7 \\
\times \underline{\ 2} =
\end{array}
$$

Multiply 8 ✗ 2 = 16. Since the next digit 7 is big, calculate
16 + 1 = 17. Write 17:

$$
\begin{array}{r}
8\ 7 \\
\times \underline{\ \ 2} = \\
1\ 7
\end{array}
$$

Multiply 7 ✗ 2 = 14. Since it is the last digit, write the unit 4 of 14:

$$
\begin{array}{r}
8\ 7 \\
\times \underline{\ \ 2} = \\
1\ 7\ 4
\end{array}
$$

Note: Performing calculations from left to right enables prompt evaluation of the result's magnitude as you instantly grasp the value of the most significant figures. This proves advantageous in daily life when employing mental calculations.

Exercises

4.1 Double the following numbers by proceeding from left to right:

a)

4 3	3 8	7 9	9 7
✗ __2 =	✗ _2 =	✗ _2 =	✗ _2 =

b)

2 4 1	4 5 6	6 8 3
✗ ____2 =	✗ ____2 =	✗ ____2 =

c)

3 2 4 5	2 4 3 7	7 6 7 9
✗ _____2 =	✗ _____2 =	✗ _____2 =

Shortcut 14 –
Doubling with expansion in addends

(Useful for mental computation)

Expand the value as a sum with multiple addends, double each addend, and sum the results.

Example:

$$342 \times 2 =$$
$$300 \times 2 \;+\; 40 \times 2 \;+\; 2 \times 2 =$$
$$600 + 80 + 4 = 684$$

Example:

$$3649 \times 2 =$$
$$3000 \times 2 \;+\; 600 \times 2 \;+\; 40 \times 2 \;+\; 9 \times 2 =$$
$$6000 + 1200 + 80 + 18 =$$
$$7200 + 98 = 7298$$

Exercises

4.2 Expand as sum of addends and double:

a) $231 \times 2 =$ $329 \times 2 =$ $263 \times 2 =$ $569 \times 2 =$

b) $3223 \times 2 =$ $4321 \times 2 =$ $2276 \times 2 =$ $6527 \times 2 =$

Shortcut 15 –
Doubling by parts

(Useful for mental computation)

This method is advantageous for numbers with many digits.

Example:

$$1\,9\,3\,4$$
$$\times \underline{\qquad 2} =$$

To prevent carries, place one or more vertical lines between the digits, ideally to the left of a tiny digit:

$$1\,9 \mid 3\,4$$
$$\times \underline{\hspace{2cm} 2} =$$

Double the individual parts:

$$1\,9 \mid 3\,4$$
$$\times \underline{\hspace{2cm} 2} =$$
$$3\,8 \quad 6\,8$$

Example:

$$2\,5\,1\,3\,5$$
$$\times \underline{\hspace{2cm} 2} =$$

$$2\,5 \mid 1 \mid 3\,5$$
$$\times \underline{\hspace{2cm} 2} =$$
$$5\,0 \quad 2 \quad 7\,0$$

Example:

$$4\,5\,3\,3\,9\,5$$
$$\times \underline{\hspace{2cm} 2} =$$

$$4\,5 \mid 3\,3\,9 \mid 5$$
$$\times \underline{\hspace{2cm} 2} =$$
$$9\,0 \quad 6\,7\,8 \quad {}^{1}0$$

$$906790$$

> By isolating the units (5), the outcome accounts for a single digit. Having obtained two digits (10), a carry of 1 occurs.

Exercises

4.3 Separate the digits and double:

a)
$$2\ 3\ 5$$
$$\times \underline{\hspace{1cm} 2} =$$

$$3\ 2\ 4$$
$$\times \underline{\hspace{1cm} 2} =$$

$$7\ 3\ 2$$
$$\times \underline{\hspace{1cm} 2} =$$

$$9\ 5\ 4$$
$$\times \underline{\hspace{1cm} 2} =$$

b)
$$4\ 5\ 1$$
$$\times \underline{\hspace{1cm} 2} =$$

$$6\ 2\ 3$$
$$\times \underline{\hspace{1cm} 2} =$$

$$5\ 8\ 2$$
$$\times \underline{\hspace{1cm} 2} =$$

c)
$$3\ 6\ 4\ 1$$
$$\times \underline{\hspace{1cm} 2} =$$

$$1\ 2\ 6\ 3\ 4$$
$$\times \underline{\hspace{1cm} 2} =$$

$$6\ 3\ 1\ 4\ 4$$
$$\times \underline{\hspace{1cm} 2} =$$

Shortcut 16 –
Multiplying and dividing by 4 and by 8

(Useful for mental computation)

To multiply by **4, double twice**.

Example:
$$24 \times 4 = 24 \times 2 \times 2 = 48 \times 2 = 96$$

Example:
$$436 \times 4 = 436 \times 2 \times 2 = 872 \times 2 = 1744$$

To multiply by **8, duplicate three times**.

Example:
$$24 \times 8 = 24 \times 2 \times 2 \times 2 = 48 \times 2 \times 2 = 96 \times 2 = 192$$

Example:
$$518 \times 8 = 518 \times 2 \times 2 \times 2 = 1036 \times 2 \times 2 = 2072 \times 2 = 4144$$

To **divide** by **4, halve twice..**

Examples:
$$48 \div 4 = 48 \div 2 \div 2 = 24 \div 2 = 12$$

$$35 \div 4 = 35 \div 2 \div 2 = 17{,}5 \div 2 =$$
$$= 17 \div 2 + 0{,}5 \div 2 = 8{,}5 + 0{,}25 = 8{,}75$$

To **divide** by **8, halve three times**.

Examples:
$$512 \div 8 = 512 \div 2 \div 2 \div 2 = 256 \div 2 \div 2 = 128 \div 2 = 64$$

$$44 \div 8 = 44 \div 2 \div 2 \div 2 = 22 \div 2 \div 2 = 11 \div 2 = 5{,}5$$

Exercises

4.4 Multiply or divide by doubling or halving several times.

a) 42 X 4 =   53 X 4 =   24 X 4 =   86 X 4 =

b) 244 ÷ 4 =   328 ÷ 4 =   454 ÷ 4 =   286 ÷ 4 =

c) 321 X 8 =   219 X 8 =   144 ÷ 8 =   1296 ÷ 8 =

Shortcut 17 –
Halving with expansion into addends

(Useful for mental computation)

Finding half of a number is equivalent to dividing it by two.
The number is converted into a sum of halved addends.

Examples:
$$468 \div 2 =$$
In this simple case, you can divide each digit by two because they are all even.

$$4 \div 2 \quad 6 \div 2 \quad 8 \div 2$$
$$2 \qquad 3 \qquad 4$$
$$234$$

Or it can be broken down as a sum of digits:
$$400 \div 2 \ + \ 60 \div 2 \ + \ 8 \div 2 =$$
$$200 + 30 + 4 = 234$$

The first method is faster and should be applied if there are even digits.
The second method is easier when there are odd digits.

Example:
$$654 \div 2 =$$
$$600 \div 2 \ + \ 50 \div 2 \ + \ 4 \div 2 =$$
$$300 + 25 + 2 = 327$$

Example:

$$753 \div 2 =$$
$$700 \div 2 \ + \ 50 \div 2 \ + \ 3 \div 2 =$$
$$350 + 25 + 1{,}5 = 376{,}5$$

Exercises

4.5 Halve by expansion in addends.

a) $42 \div 2 =$ $56 \div 2 =$ $74 \div 2 =$ $87 \div 2 =$

b) $286 \div 2 =$ $353 \div 2 =$ $471 \div 2 =$ $936 \div 2 =$

c) $1326 \div 2 =$ $3316 \div 2 =$ $5147 \div 2 =$ $2273 \div 2 =$

Shortcut 18 –
Halving on a single line

This approach is helpful when dealing with long numbers that contain odd digits, which can call for the usage of carry.

Example:

$$4\,5\,8 \div 2 =$$

Starting from the left and proceeding to the right we get:

$4 \div 2 = 2$. Write 2.

$$4\,5\,8 \div 2 = \mathbf{2}$$

$5 \div 2 = 2$ with remainder $r = 1$. Write 2 and report the remainder beside the next digit of the dividend as a small number to the right, slightly down:

$$4\,5\,{}_1 8 \div 2 = \mathbf{22}$$

The remainder 1, with 8 gives 18. Therefore:
$18 \div 2 = 9$. Write 9.

$$4\,5\,{}_1 8 \div 2 = \mathbf{229}$$

Example:
$$7\,5\,2\,6\,3 \div 2 =$$
Starting from the left:

$7 \div 2 = \mathbf{3}\,r\,1$ $7\,_15\,2\,6\,3 \div 2 = \mathbf{3}$

$15 \div 2 = \mathbf{7}\,r\,1$ $7\,_15\,_12\,6\,3 \div 2 = 37$

$12 \div 2 = \mathbf{6}$ $7\,_15\,_12\,6\,3 \div 2 = 376$

$6 \div 2 = \mathbf{3}$ $7\,_15\,_12\,6\,3 \div 2 = 3763$

$3 \div 2 = \mathbf{1}\,r\,\mathbf{1}$ $7\,_15\,_12\,6\,3 \div 2 = 37631$ resto $= 1$

Since you have 1 of remainder and $1 \div 2 = 0{,}5$ you can write the result as:

37631.5

Example:
$$1\,7\,4\,3 \div 2 =$$

$1 \div 2 = 0\,r\,1$ $1\,_17\,4\,3 \div 2 = 0$

$17 \div 2 = \mathbf{8}\,r\,1$ $1\,_17\,_14\,3 \div 2 = 08$

$14 \div 2 = \mathbf{7}$ $1\,_17\,_14\,3 \div 2 = 087$

$3 \div 2 = \mathbf{1}\,r\,1$ $1\,_17\,_14\,3 \div 2 = 0871\,r\,1 = \mathbf{871.5}$

Exercises

4.6 Halve on a single line.

a) $568 \div 2 =$ $736 \div 2 =$ $818 \div 2 =$ $327 \div 2 =$

b) $3672 \div 2 =$ $4726 \div 2 =$ $5137 \div 2 =$ $8765 \div 2 =$

c) $36912 \div 2 =$ $89576 \div 2 =$ $733752 \div 2 =$ $854771 \div 2 =$

Shortcut 19 –
Halving and doubling

(Useful for mental computation)

This shortcut proves useful when multiplying two numbers where at least one is even.

Example:
$$18 \times 9 =$$
18 is even: you can halve 18 and multiply by 2:
$$9 \times 2 \times 9 =$$
Multiply 9 by 9
$$81 \times 2 = 162$$

In this case, the multiplication changed from
2 digits x 2 digits to 2 digits x 1 digit

Example:
$$28 \times 12 =$$

You can halve both digits and multiply by two twice:

$14 \times 2 \times 6 \times 2 =$	*Halve 28 e 12*
$14 \times 6 \times 2 \times 2 =$	*Multiply 14 by 6*
$84 \times 2 \times 2 =$	*Double 84*
$84 \times 2 = 168$	*Double 168*
$168 \times 2 = \mathbf{336}$	

Example:
$$125 \times 22 =$$

$125 \times 2 \times 22 \div 2 =$	*Double 125 and halve 22*
$250 \times 11 =$	*Write 250 as 25 x 10*
$\underline{25} \times 10 \times \underline{11} =$	*Multiply 25 by 11*
$275 \times 10 = \mathbf{2750}$	*Multiply by 10*

Example:

$$3.5 \times 18 =$$

$3.5 \times 2 \times 18 \div 2 =$	*Double 3.5 and halve 18*
$7 \times 9 = 63$	*Multiply 7 by 9*

Exercises

4.7 Multiply by halving and doubling appropriately.

a) $24 \times 5 =$ $32 \times 22 =$ $45 \times 16 =$ $16 \times 14 =$ $18 \times 35 =$

b) $248 \times 2,5 =$ $125 \times 24 =$ $48 \times 1,5 =$ $180 \times 3,5 =$ $64 \times 25 =$

Shortcut 20 –
Multiplying by 5, 15, 25, 50, 0,5

(Useful for mental computation)

Applying this method involves multiplying by 10, 100, etc., and then halving one or more times.

Multiplying by 5 is equivalent to multiplying by 10 and dividing by 2

Example:

$$32 \times 5 =$$
$$32 \times 10 \div 2 =$$
$$320 \div 2 = \mathbf{160}$$

Multiplying by 50 is equivalent to multiplying by 100 and dividing by 2.

Example:

$$54 \times 50 =$$

$$54 \times 100 \div 2 =$$
$$5400 \div 2 = \mathbf{2700}$$

Multiplying by 15 is equivalent to resolving ten times the number and adding half of that value:

Example:

$$12 \times 15 =$$
$$12 \times 10 + (12 \times 10 \div 2)$$
$$120 + 60 = \mathbf{180}$$

Multiplying by 25 is equivalent to multiplying by 100 and dividing by 4:

Example:

$$43 \times 25 =$$
$$43 \times 100 \div 4$$
$$4300 \div 4 =$$
(Halve 2 times 4000 and 300).
$$\mathbf{1075}$$

Multiplying by 0,5 is equivalent to dividing by 2:

Example:

$$28 \times 0,5 =$$
$$28 \div 2 = \mathbf{14}$$

Exercises

4.8 Solve the following multiplications:

a) $32 \times 5 =$ $17 \times 5 =$ $422 \times 5 =$ $165 \times 5 =$ $318 \times 5 =$

b) $28 \times 50 =$ $125 \times 50 =$ $312 \times 50 =$ $182 \times 50 =$ $64 \times 50 =$

c) $64 \times 15 =$ $18 \times 15 =$ $220 \times 15 =$ $24 \times 25 =$ $72 \times 25 =$

d) $128 \times 25 =$ $325 \times 25 =$ $412 \times 0,5 =$ $182 \times 0,5 =$ $256 \times 0,5 =$

Shortcut 21 –
Dividing by 5, 25, 50, 0.5

(Useful for mental computation)

This strategy is applied by doubling and dividing by a power of 10.

Example:
$$28 \div 5 =$$
Since $5 = 10 \div 2$, we double and divide by 10, so we get:

$$28 \div (10 \div 2) =$$
$$28 \div 10/2 =$$
$$28 \times 2 /10$$
$$\mathbf{28 \times 2 \div 10 =}$$
$$56 \div 10 = \mathbf{5.6}$$

> **Division by 5:**
> **Double and divide by 10**

Example:
$$123 \div 5 =$$
$$123 \times 2 \div 10 =$$
$$246 \div 10 = \mathbf{24.6}$$

Example:
Dividing by 25 means dividing by 100/4, that is, multiplying by 4/100.

$$118 \div 25 =$$
$$\mathbf{118 \times 4 \div 100 =}$$
$$472 \div 100 = \mathbf{4.72}$$

> **Division by 25:**
> **Multiply by 4, divide by 100**

Examples:
Dividing by 50 means dividing by 100/2, that is, multiplying by 2/100.

$$147 \div 50 =$$
$$\mathbf{147 \times 2 \div 100 =}$$
$$294 \div 100 = \mathbf{2.94}$$

> **Division by 50:**
> **Multiply by 2, divide by 100**

$$2215 \div 50 =$$
$$2215 \times 2 \div 100 =$$
$$4430 \div 100 = \mathbf{44.3}$$

Example:

Dividing by 0.5 means dividing by 1/2, that is, multiplying by 2.

$$318 \div 0.5 =$$

$$318 \times 2 = \mathbf{636}$$

> **Division by 0.5:**
> **Doubling**

Exercises

4.9 Solve the following divisions:

a) $42 \div 5 =$ $88 \div 5 =$ $321 \div 5 =$ $315 \div 25 =$ $618 \div 25 =$

b) $91 \div 25 =$ $125 \div 25 =$ $62 \div 50 =$ $142 \div 50 =$ $111 \div 50 =$

c) $32 \div 0.5 =$ $61 \div 0.5 =$ $224 \div 0.5 =$ $442 \div 0.5 =$ $99 \div 0.5 =$

5. FAST MULTIPLICATIONS

Shortcut 22 –
Multiplying two-digit numbers with tens equal
and with complementary units

(Useful for mental computation)

When you need to multiply two two-digit numbers with equal tens digits, and the sum of their units digits is ten, a shortcut is employed for obtaining the result through mental calculation alone.
The rule is, "*For one more than the previous one*".

Example:

$$34 \times 36$$

First verification:
The figures of tens are worth 3 for both factors.

Second verification:
The sum of the units is worth 10:   4 + 6 = 10.

Multiply the ten (3) by itself and add 1, following the rule "For *one more than the previous one*".

$$3 \times (3 + 1) = 3 \times 4 = \mathbf{12}$$

Thus, you obtain the left-hand side of the result.
To get the right-hand side, multiply the units:

$$4 \times 6 = \mathbf{24}$$

$$
\begin{array}{ll}
\overline{} & \\
|| & \textit{3 x (3 + 1) = 12} \\
3\,4 \;\times\; 3\,6 = 1224 & \\
|\underline{}| & \textit{4 x 6 = 24}
\end{array}
$$

So you get: **1224**.

Example:

$$78 \times 72$$

Verified that the tens are equal and the units complementary, you can proceed with the shortcut:

$$7 \times (7 + 1) \;=\; 7 \times 8 = \; \mathbf{56}$$
$$8 \times 2 = \mathbf{16}$$
$$78 \times 72 = \mathbf{5616}$$

You can apply this pattern to find the squares of two-digit numbers ending in 5, in which case all results will end in 25::

Squares of two-digit numbers ending with 5

This is a special case which use the same shortcut. *(Multiply two-digit numbers with tens equal and with complementary units).*

Example:

$$35^2 = 35 \times 35$$

Tens are equal, and units too, equal to 5, and therefore complementary:

$$3 \times (3 + 1) \;=\; 3 \times 4 = \; \mathbf{12}$$
$$5 \times 5 = \mathbf{25}$$
$$35^2 = \mathbf{1225}$$

Example:

$$65^2 = 65 \times 65$$
$$6 \times 7 = 42$$
$$5 \times 5 = 25$$
$$65 \times 65 = 4225$$

You can apply this pattern, in the easy cases, with numbers with three digits:

$$115^2 = 115 \times 115$$
$$11 \times 12 = 132$$
$$115 \times 115 = 13225$$

Exercises

5.1 Check the conditions and multiply in mind with the rule "*For one more than the previous*".

a) 45 × 45 =   32 × 38 =   83 × 87 =   95 × 95 =

b) 18 × 12 =   66 × 64 =   49 × 41 =   37 × 33 =

c) 15^2 = 45^2 = 65^2 = 75^2 = 115^2 =

d) 105 × 105 =   102 × 108 =   125 × 125 =   124 × 126 =

Shortcut 23 –
Multiplying two-digit numbers with units equal and with complementary tens

(Useful for mental computation)

When multiplying two two-digit numbers with equal unit digits and a sum of ten in their tens digits, a shortcut can be employed for mental calculation to obtain the result.

Example:

$$72 \times 32$$

First verification:
The figure of units is worth 2 for both factors.
Second verification:
The sum of the tens is worth 10: $7 + 3 = 10$.
The pattern is:
To get the left-hand side of the result, we multiply the tens and sum the unit:

$$7 \times 3 + 2 = 23$$

To get the right-hand side of the result, multiply the units:

$$2 \times 2 = 04$$

$$
\begin{array}{ll}
|\overline{}^{\times}|^{+}| & 7 \times 3 + 2 = 23 \\
7\,2 \times 3\,2 = \mathbf{2304} & \\
|\underline{}_{\times}\underline{}| & 2 \times 2 = 04
\end{array}
$$

So you get: **2304**.

Example:

$$84 \times 24$$

Verified that the tens are complementary and the units equal, the shortcut can be proceeded with:

$$8 \times 2 + 4 = \mathbf{20}$$
$$4 \times 4 = \mathbf{16}$$
$$84 \times 24 = \mathbf{2016}$$

Exercises

5.2 Check the conditions and multiply.

a) $65 \times 45 =$ $23 \times 83 =$ $38 \times 78 =$ $59 \times 59 =$

b) $81 \times 21 =$ $66 \times 46 =$ $94 \times 14 =$ $73 \times 33 =$

c) $89 \times 29 =$ $41 \times 61 =$ $63 \times 43 =$ $13 \times 93 =$

Shortcut 24 –
Multiplying two-digit numbers with units equal to 5

(Useful for mental computation)

In two-digit multiplications ending in 5, when no cases of complementary digits between tens or units are observed, another pattern can be applied:

Example:

$$25 \times 65$$

Multiply the tens:

$$2 \times 6 = 12$$

Add up the tens and divide by 2:

$$\frac{2+6}{2}=4$$

(The sum of the tens is even, so the semi-sum is integer).

To get the left-hand side of the solution, add up the results:

$$12 + 4 = \textbf{16}$$

To get the right-hand side, multiply the units:

$$5 \times 5 = \textbf{25}$$

$$25 \times 65 = \textbf{1625}$$

Example:

$$35 \times 65$$

Multiply the tens:

$$3 \times 6 = 18$$

Add up the tens and divide by 2:

$$\frac{3+6}{2}=4,5$$

(The sum of the tens is odd, so the semi-sum has a decimal).

To get the left-hand side of the solution, add up the results:

$$18 + 4,5 = 22,5$$

The left part we have obtained represents the hundreds of the result, and the decimal part of 22.5, that is 0.5, are tens. So we have 22 hundred and five tens.

The result on the right-hand side will be 5 x 5 = 25 and must be augmented by the five tens obtained from the preceding calculation:

$$\overset{5}{22\ 25}$$

we get:

$$35 \times 65 = \textbf{2275}$$

Summing up:

The **left-hand side of the result** is obtained by **adding the product and the semi-sum of the tens**. In case a decimal (0.5) is obtained (the sum of the tens is odd), the **integer part** *(to the left of the decimal point) is considered*.

In the event that **no decimal was obtained** (the sum of the tens is even), the **right-hand side** of the result will be **25**. **If a decimal component (0.5) was found** in the previous computation, it will instead be **75** because the sum of the tens is odd.

Example:

The sum of tens is **even**:

$$15 \times 75$$

$$1 \times 7 = 7 \qquad\qquad \frac{1 + 7}{2} = 4$$

> $1 + 7 = 8$
> is even.
> The result ends
> with 25

$$7 + 4 = 11$$
$$15 \times 75 = \mathbf{1125}$$

Example:

The sum of tens is **odd**:

$$25 \times 75$$

$$2 \times 7 = 14 \qquad\qquad \frac{2 + 7}{2} = 4{,}5$$

> $2 + 7 = 9$
> is odd.
> The result ends with
> 75

$$14 + 4 = 18$$
$$15 \times 75 = \mathbf{1875}$$

Exercises

5.3 Calculate in mind:

a) $35 \times 75 =$ $\qquad$ $25 \times 95 =$ $\qquad$ $35 \times 45 =$ $\qquad$ $25 \times 35 =$

b) $95 \times 15 =$ $\qquad$ $85 \times 25 =$ $\qquad$ $65 \times 35 =$

c) $95 \times 25 =$ $\qquad$ $75 \times 35 =$ $\qquad$ $85 \times 55 =$

Shortcut 25 –
Numbers close to 10, 100, 1000 ecc.

(Useful for mental computation)

When performing multiplication, the two factors can have values close to a power of ten.

For example, 8 x 9 has factors close to 10; 89 x 97 has factors close to 100; 1005 x 998 has factors close to 1000.

A specific calculation method exploits the surplus or shortfall of factors with respect to a reference base, typically a power of 10. This fact enables the easy mental computation of products, even when dealing with large numbers.

Two rules are applied: "*By deficiency*" e "*All from nine, the last from 10*".

Multiplication under a base

Example:

$$7 \times 8 =$$

The reference base is 10.

The factor 7 is three short of 10, while the factor 8 is two short of 10.

$$7 \quad [- 3 \text{ compared to } 10]$$

$$\times\,\underline{8} \quad [- 2 \text{ compared to } 10]$$

The outcome is calculated in two steps:

First, we find the **left-hand side** of the result by **subtracting from one of the two factors the deficiency of the other one** concerning the power of 10.

(Negative figures are indicated by bar numbers and represent deficiencies, that is, how much is missing to get the base, in this example, 10).

$$7 \ [\overline{3}]$$

$$x \ \underline{8} \ [\overline{2}]$$

(cross-wise operation)

$$7 - 2 = 5, \text{ oppure } 8 - 3 = 5$$

Then, we find the **right-hand side** of the result by **multiplying the two deficiencies**.

$$\overline{3} \times \overline{2} = 6$$

We get:

$$7 \ [\overline{3}$$
$$x \ \underline{8} \ [\overline{2}]$$

$$5 \ | \ 6$$

So the result is: **56**

Note: Multiplying two negative deficiencies yields a positive result.

Example:

$$98 \times 89$$

The reference base is 100.

$$98 \ [-2 \text{ compared to } 100]$$
$$x \ \underline{89} \ [-11 \text{ compared to } 100]$$

First, we find the **left-hand side** of the result by **subtracting from one of the two factors the deficiency of the other one** concerning the power of 10.

$$98 \ [0\overline{2}]$$
$$x \ \underline{89} \ [\overline{11}]$$

(cross-wise operation)

$$98 - 11 = 87, \text{ or } 89 - 02 = 87$$

Then, we find the **right-hand side** of the result by **multiplying the two deficiencies**. The right-hand side has available two digits because the base is 100.

$$\overline{2} \times \overline{11} = 22$$

$$
\begin{array}{r}
98 \; [\overline{02}] \\
\times\, \underline{89} \; [\overline{11}] \\
87 \mid 22
\end{array}
$$

So the result is: **8722.**

Example:

$$
\begin{array}{r}
997 \; [\overline{003}] \\
\times\, \underline{992} \; [\overline{008}]
\end{array}
$$

$$992 - 003 = \mathbf{899}; \quad \overline{003} \times \overline{008} = \mathbf{024}$$
$$997 \times 992 = \mathbf{899024}$$

Note: *The units of the result have as many digits reserved for as there are zeros in the reference base.*

Exercises

5.4 Multiply.

a) 97 x 99 =   86 x 98 =   91 x 89 =   88 x 95 =

b) 92 x 81 =   89 x 75 =   87 x 97 =   84 x 98 =

c) 988 x 994 =   985 x 996 =   998 x 875 =   989 x 888 =

> *Multiplication over a base*

Example:
$$12 \times 13 =$$

The reference base is 10.

$$\mathbf{12} \; [+\, 2 \text{ over } 10]$$

$$\times\, \underline{\mathbf{13}} \; [+\, 3 \text{ over } 10]$$

First, we find the **left-hand side** of the result by **adding to one of the two factors the excess of the other one** concerning the power of 10.

In this example, 12 is more than 2 over 10, while 13 is in excess of 3.

12 [2]

x 13 [3]

(cross-wise operation)

12 + 3 = **15**, or 13 + 2 = **15**

Then, we find the **right-hand side** of the result by **multiplying the two excesses**.

2 x 3 = **6**

12 [2]

x 13 [3]

15 | 6

So the result is: **156**

Note: *Since the base considered is* **10**, *only* **one digit** *will be available on the right-hand side of the result. Therefore, if the product of excesses or defects were to be two digits, the ten of that product would go as a carry on the left side of the result.*

Example:

The reference base is 100.

105 **x** 125 =

105 [05]

x **125** [25]

(cross-wise operation)

105 + 25 = **130**, or 125 + 5 = **130**

which is the **left-hand side of the** product.

Then, we find the **right-hand side** of the result by **multiplying the two excesses**.

$$5 \times 25 = \mathbf{125}$$

Since the base is 100, you have two digits for the right-hand side of the result, so 25 is the right-hand side, and the hundred (1 of 125) goes as a carry on the left-hand side.

So the result is:

$$105 \ [05]$$

$$\times \underline{125} \ [25]$$
$$1$$
$$130 \mid 25$$
$$\mathbf{13125}$$

Exercises

5.5 Multiply.

a) 102 x 104 =   111 x 103 =   1001 x 1009 =   120 x 115 =

b) 111 x 106 =   1012 x 1011 =   113 x 103 =   1125 x 1002 =

c) 135 x 111 =   123 x 105 =   1250 x 1011 =   1012 x 1106 =

Multiplication one factor above and one factor below a base

Example:

$$103 \times 96 =$$

The first factor is above 3 out of 100, while the second factor is below 4 out of 100.

$$103 \ [03]$$

$$\times \underline{96} \ [\overline{04}]$$
(cross-wise operation)

The left-hand side of the product is:

$$103 - 4 = 99, \text{ or } 96 + 3 = \mathbf{99}$$

The right-hand side of the result is the product between the excesses:

The right-hand side has two digits available because the base is 100.

(The product between a positive and a negative number is negative).

$$3 \times \overline{4} = \overline{12}$$

So the result is:

$$103 \quad [03]$$

$$\times \quad \underline{96} \quad [\overline{04}]$$

$$99 \mid \overline{12}$$

$$99\overline{12}$$

which, transformed from a bar number to a positive one, becomes:

$$99 \mid \overline{12}$$

$$\mathbf{9888}$$

(9900 – 12 = 9888 or, using the rules **"All from nine, the last from ten"** *and* **"One less than the previous one"***: 99 minus 1 is 98, wich is the left-hand side of the result. About the right-hand side, we have 9 – 1 = 8 and 10 – 2 = 8, so 88 is the right-hand side of the result).*

Example:

$$9996 \times 10002 =$$

The first factor is down by four from 10,000, while the second is up by two from 10000.

$$9996 \quad [\overline{0004}]$$

$$\times \underline{10002} \quad [0002]$$

(cross-wise operation)

The left-hand side of the product is:

$$9996 + 2, \text{ or } 10002 - 4 = \mathbf{9998}$$

The right-hand side of the result has **four digits available**, since the **base is 10000**, and is the product between the excesses and/or deficiencies:

$$\overline{0004} \times 0002 = \mathbf{0008}$$

$$9996 \quad [\overline{0004}]$$

$$\times \underline{10002} \quad [0002]$$

$$9998 \mid \overline{0008}$$

So the result is:

$$9998\overline{0008}$$

which, transformed from a bar number to a positive one, becomes:

99979992

(Using the rules **"All from nine, the last from ten"** and **"One less than the previous one"**: 9998 minus 1 is 9997, wich is the left-hand side of the result; $9 - 0 = 9$ repeated for the three zeros, and $10 - 8 = 2$, so 9992 is the right-hand side of the result).

Exercises

5.6 Multiply.

a) $102 \times 99 =$ $111 \times 89 =$ $112 \times 88 =$ $120 \times 75 =$

b) $113 \times 96 =$ $1012 \times 998 =$ $1014 \times 998 =$ $1125 \times 996 =$

c) $105 \times 97 =$ $1115 \times 996 =$ $1250 \times 998 =$ $1012 \times 997 =$

Shortcut 26 –
Squares of numbers close to 50

Another specific method is for squares of numbers close to 50. Two steps are involved in calculating the result:

The excess (or deficit) of the number relative to 50 is obtained algebraically by adding 25 (half of 50) to the left.

The square of the surplus or shortfall is on the right.

Example:
$$53^2$$
53 exceeds 50 by 3, so 3 is added to 25, obtaining the left-hand side of the result:
$$25 + 3 = \textbf{28}$$
The right-hand side is the square of 3, so 3 x 3 = **09**. We get:
$$53^2 = \textbf{2809}$$

Example:
$$\textbf{46}^2$$
46 is down by four from 50.

The left-hand side is:
$$25 + (-4) = 25 - 4 = \textbf{21}$$

The right-hand side will be the square of the deficiency:
$$(-4)^2 = (-4) \times (-4) = \textbf{16}$$
So:
$$46^2 = \textbf{2116}$$

Exercises

5.7 Calculate the squares of the following numbers:

a) $47^2 =$ $54^2 =$ $39^2 =$ $62^2 =$ $41^2 =$

b) $38^2 =$ $53^2 =$ $44^2 =$ $56^2 =$ $43^2 =$

Shortcut 27 –
Multiplying by 999

This pattern works when the **two factors have** the **same number of digits**.

Memorize the following two Vedic aphorisms:

- *One less than the previous one*
- *All from nine, the last from ten*

Example:

$$632 \times 999 =$$

To find the left-hand side of the result, subtract one from the multiplicand (*One less than the previous one*):

$$632 - 1 = \textbf{631}$$
$$632 \times 999 = \textbf{631} \ldots$$

To find the right-hand side of the result, starting with the multiplicand 632, subtract the hundreds 6 from 9, then the tens 3 from 9. Subtract the units 2 from 10 (*All from nine, last from ten*):

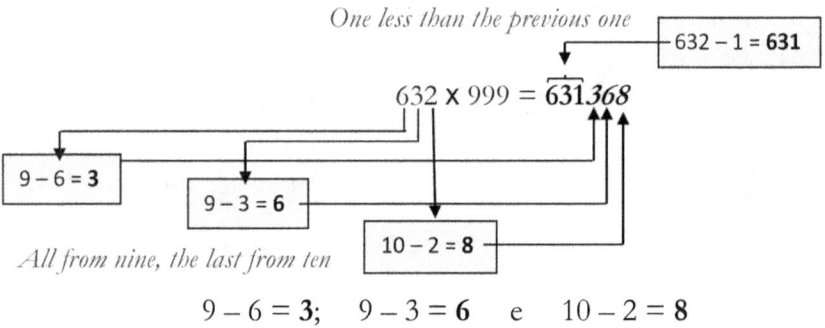

$$9 - 6 = \textbf{3}; \quad 9 - 3 = \textbf{6} \quad e \quad 10 - 2 = \textbf{8}$$

The right-hand side of the result is 368.

$$632 \times 999 = \textbf{631368}$$

(You can extend the rule to big numbers whether the two factors have the same number of digits).

Example:

$$274541 \times 999999 = \underline{274540}\,\underline{725459}$$
$$-1 \quad Complement$$

Exercises

5.8 Multiply.

a) 122 x 999 =   213 x 999 =   232 x 999 =   742 x 999 =

b) 512 x 999 =   957 x 999 =   104 x 999 =   625 x 999 =

c) 405 x 999 =   1115 x 9999 =   1250 x 9999 =   1012 x 9999 =

Shortcut 28 –
Multiplying by 101, 102, 201, 202 ecc.

Example:

$$325 \times 101 =$$

The multiplier 101 has three digits, but it could be written, adding a zero to the left, as 0101.
Thus 325 can be multiplied for each group 01. The result of each product, however, has three digits and not two, so the first digit to the left of the first result group will go as carry:

$$325 \times 01 = {}_3 25$$

> **Multiply the 1ˢᵗ group**
> $325 \times 01 = {}_3 25$
> Write 25, carry 3

$$325 \times 0101 = \dots {}_3 25$$

When you multiply 325 by the second group 01, you will then have to add the carry:

$$325 \times 01 = 325; \quad 325 + 3 = 328$$

$$325 \times \underline{0101} = 328_3 25$$

> **Multiply the 2st group**
> $325 \times 01 = 325$
> Add the carry 3:
> $325 + 3 = 328$

Result: **32825**

Example:

$$234 \times 102 =$$

Multiply 234 $\times$ 02 = *468* (First group): write 68, carry 4:

$$234 \times \underline{102} = \quad _4 68$$

Multiply 234 $\times$ 1 = *234* (second group) and add the carry:
234 + 4 = 238
Write 238

$$234 \times \underline{102} = 238_4 68$$

Result: **23868**

Example:

$$123 \times 2001 = \quad 123 \times 002001 = \mathbf{246}\mathbf{\textit{123}}$$

(Right-end side: 123 $\times$ 001 = 123; Left-hand side: 123 $\times$ 002 = 246)

Exercises

5.9 Multiply.

a) 81 $\times$ 101 =   27 $\times$ 101 =   32 $\times$ 101 =   74 $\times$ 101 =

b) 52 $\times$ 102 =   24 $\times$ 201 =   36 $\times$ 202 =   41 $\times$ 102 =

c) 412 $\times$ 101 =   315 $\times$ 102 =   125 $\times$ 202 =   11 $\times$ 103 =

Shortcut 29–
Cross-wise multiplication

Two digits by two digits

Example:

$$3\,2$$
$$\times\ \underline{1\,4} =$$

Start by **vertically** multiplying the two units: $4 \times 2 = 8$.

Starting from the right, multiply 4×2 and write the result 8 in the units column on the right:

$$3\,2$$
$$|$$
$$\times\ \underline{1\,4} =$$
$$\mathbf{8}$$

Multiply **diagonally**: units by tens and tens by units by adding up the results:

$$4 \times 3 + 1 \times 2 = 12 + 2 = \mathbf{14}$$

Write 4 in the tens column and carry 1.

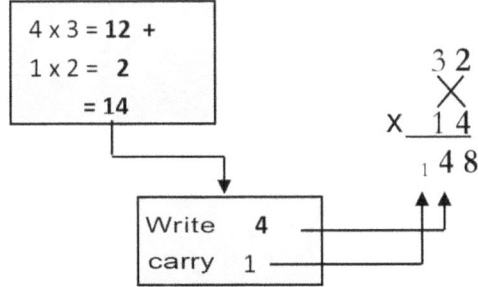

Finally, multiply, **vertically**, the two tens: $1 \times 3 = 3$.

Add the carry: $3 + 1 = \mathbf{4}$ and write the result on the left-hand side:

$$3\,2$$
$$|$$
$$\times\ \underline{1\,4} =$$
$$\mathbf{4}\,_1\mathbf{4}\,\mathbf{8} \qquad \text{Result: } \mathbf{448}$$

Exercises

5.10 Multiply with a cross-wise pattern.

a) 56   65   35   33   88   61

  x _13_ =   **x** _11_ =   **x** _35_ =   **x** _37_ =   **x** _14_ =   **x** _69_ =

b) 25   56   76   94   91   41

  x _85_ =   **x** _56_ =   **x** _36_ =   **x** _98_ =   **x** _89_ =   **x** _59_ =

c) 74   85   42   69   51   76

  x _45_ =   **x** _47_ =   **x** _49_ =   **x** _27_ =   **x** _64_ =   **x** _79_ =

Three digits by three digits

Example:

$$3\,2\,1 \times$$
$$\underline{1\,2\,5} =$$

$3\,2\,1$ $(1 \times 5 = 5)$
| *(vertical multiplication on the right-hand column)*
$\times \underline{1\,2\,5} =$
5

$3\,2\,1$ $(2 \times 5 + 1 \times 2 = 10 + 2 = 12)$ 2, carry 1
$\times$ *(diagonal multiplication on two columns on the right)*
$\times \underline{1\,2\,5} =$
$_1 2\,5$

$\boxed{\text{Carry}}$

$3\,2\,1$ $(5 \times 3 + 2 \times 2 + 1 \times 1 = 15 + 4 + 1 = 20;$ $20 + 1 = 21)$
$\times$ *(vertical - diagonal multiplication)*   1, carry 2
$\times \underline{1\,2\,5} =$
$_2 1_1 2\,5$

$\boxed{\text{Carry}}$

$3\,2\,1$ $(2 \times 3 + 2 \times 1 = 6 + 2 = 8;$ $8 + 2 = 10)$ 0, carry 1
$\times$ *(diagonal multiplication on two columns to the left)*
$\times \underline{1\,2\,5} =$
$_1 0_2 1_1 2\,5$

$\boxed{\text{Carry}}$

$3\,2\,1 \times$ $(3 \times 1 = 3;$ $3 + 1 = 4)$
| *(vertical multiplication on the left-hand column)*
$\times \underline{1\,2\,5} =$
$4_1 0_2 1_1 2\,5$ Result: **40125**

Note: *To multiply three figures by two figures, enter a zero on the left-hand side of the two-digit number. Example:*

$$3\,2\,1 \times$$
$$\underline{0\,2\,5} =$$
$$0\,8_2 0_1 2\,5$$

Result: **8025**

Exercises:

5.11 Multiply with a cross-wise pattern.

a) 522   234   311   432   637   668

X _243_ = X _321_ =   X _365_ =   X _556_ =   X _326_ =   X _122_ =

b) 761   823   541   921   680   415

X _151_ = X _271_ =   X _123_ =   X _285_ =   X _652_ =   X _325_ =

c) 216   728   336   534   636   837

X _335_ = X _277_ =   X _492_ =   X _437_ =   X _052_ =   X _025_ =

Shortcut 30 –
Squares of two- and three-digit numbers

Duplex

The **Duplex** of a number is one of the fundamental tools of Vedic Mathematics used to calculate squares.

The Duplex D of a number is calculated as follows:

If the number consists of
- **one figure**, → just calculate its square: multiply the number by itself.

  Example:
  The Duplex of 8 is 8 X 8 = 64. It can be denoted as D(8) = 64.

$$D(8) = 64 \qquad \boxed{8 \times 8 = 64}$$

- **two figures,** → calculate the double product between the figures.

Example:

The Duplex of 24 is 2 × (2 × 4) = 16, and it is write as
D(24) = 16.

$$D(24) = 16 \qquad \boxed{2 \times (2 \times 4) = 16}$$

- **three figures,** → calculate the Duplex of the first and the last figures, then add the Duplex (square) of the middle.

Example:

The Duplex of 325 is 2 × (3 × 5) + 2 × 2 = 30 + 4 = 34.
So: D(325) = 34

$$D(3\ 2\ 5) = 34$$

$$\boxed{2 \times (3 \times 5) + 2 \times 2 = 34}$$

Examples:

D(3) = 3 × 3 = 9;
D(46) = 2 × (4 × 6) = 48;
D(347) = 2 × (3 × 7) + 4 × 4 = 42 + 16 = 58

Squares

The square of 26 is worth 26 × 26 and is written 26^2 (a small two is placed in the upper right corner of the number).

$$
\begin{array}{r}
26 \\
\times\ \underline{26} = \\
676
\end{array}
$$

The result is immediate by the multiplication tables when the number consists of a single digit.

Equally simple is if the digit is followed by one or more zeros: you multiply this by itself and add twice as many zeros.

Examples:

$$40^2 = [4 \times 4][00] = 1600$$

$$700^2 = [7 \times 7][0000] = 490000$$

Use the Duplex when calculating the square by the method of Vedic mathematics.

Example:

$$23^2$$

Calculating from right to left:

- Square of the unit: (D3) = 3 x 3 = **9**

$$23^2 = \ldots\ldots 9 \qquad \boxed{D(3) = 9}$$

- Double product of the two figures: D(23) = 2 x (2 x 3) = 12
Since 12 consists of two digits, write 2 (unit of **12**) with a carry of 1.

$$23^2 = \ldots \,{}^1 29 \quad \boxed{D(23) = 2 \times (2 \times 3) = 12}$$

- Square of the ten: D(2) = 2 x 2 = 4.

Add the carry: 4 + 1 = **5**

$$23^2 = \quad 5^1 29 = \mathbf{529} \qquad \boxed{\begin{array}{l} D(2) + 1 = 2 \times 2 + 1 = \\ 4 + 1 = 5 \end{array}}$$

Examples:

$$123^2$$

By calculating from right to left:

- Square of the units: (D3) = 3 x 3 = **9**. Write **9**. $\qquad \boxed{D(3) = 9}$

1 2 *3* $\qquad$ D(3) $\rightarrow 3^2 \rightarrow \ldots\ldots$ **9**

- Double product of the two figures on the right-hand side: D(23) = 2 x (2 x 3) = 12;
whrite **2**, carry 1.

1 **23** $D(23) \rightarrow 2 \times (2 \times 3) \rightarrow ...^1 2\,9$ $\boxed{D(23) = 2 \times (2 \times 3) = 12}$

- Double product of the first and last figures + square of the one in the middle:

$2 \times (1 \times 3) + 2^2 = 10$. Add the carry: $10 + 1 = \mathbf{11}$.

Write **1**, carry 1.

123 $D(123) \rightarrow 2 \times (1 \times 3) + 2^2 + 1 \rightarrow ...^1\mathbf{1}\,^1 2\,9$

- Double product of the two figures on the left-hand side: $D(12) = 2 \times (1 \times 2) = 4$
Add the carry: $4 + 1 = \mathbf{5}$. Write **5**.

123 $D(12) \rightarrow 2 \times (1 \times 2) + 1 \rightarrow ...\mathbf{5}^1 1\,^1 2\,9$

- Square of the left-hand side figure. $D(1) = 1 \times 1 = \mathbf{1}$. Write **1**.

123 $D(1) \rightarrow 1^2 \rightarrow \mathbf{1}\,5^1 1\,^1 2\,9$

$$123^2 = \mathbf{15129}$$

Exercises:

5.12 Calculate the following squares by the duplex pattern:

a) $7^2 =$ $23^2 =$ $46^2 =$ $63^2 =$ $87^2 =$ $68^2 =$

b) $125^2 =$ $233^2 =$ $341^2 =$ $443^2 =$ $628^2 =$ $992^2 =$

6. FAST DIVISIONS

Shortcut 31 –
Separating and dividing

(Useful for mental computation)

You can apply this pattern when, by appropriately separating the digits of the dividend, you obtain numbers that are multiples of the divisor:

Example

$$486 \div 2$$

Separating the figures of the dividend gives:

$$4 \mid 8 \mid 6 \div 2$$

Since each digit is a multiple of the divisor, divide each of these by 2, getting 243.

The quantity of figures in every section of the outcome matches the amount of digits in every section of the dividend.

Example

$$18612 \div 6$$
$$18 \mid 6 \mid 12 \div 6$$
$$03 \mid 1 \mid 02 = 3102$$

Example

$$2212133 \div 11$$
$$22 | 121 | 33 \div 11$$
$$02 | 011 | 03 = 201103$$

Exercises:

6.1 Calculate by directly writing down the result.

a) $71428 \div 7 =$ $\qquad$ $152535 \div 5 =$ $\qquad$ $27549 \div 9 =$

b) $42618 \div 6 =$ $\qquad$ $361624 \div 4 =$ $\qquad$ $85616 \div 8 =$

c) $1155121 \div 11 =$ $\qquad$ $2414436 \div 12 =$ $\qquad$ $304560 \div 15 =$

d) An Australian entrepreneur owns 33126 sheep to bequeath, in equal parts, to his three sons. How many sheep will each son inherit?

Shortcut 32 –
Divisions close to a base

Division under a base

When a divisor is a number close to a base power of 10 (*e.g., 10, 100, 1000, etc.*), Vedic Mathematics uses the complement to the base to simplify and speed up the calculation.

For this reason, the rule *"All from 9 and the last from 10"* applies.

Example:

Dividing a number with a divisor next to and less than a base:

$$1512 \div 89$$

89 is close to 100, and its complement is $100 - 89 = 11$.

Place 89 to the left of the pattern and 1512 to the right. Then, separate two figures from the right since the base is 100.

The remainder of the division will have at least two digits.

Under the divisor 89, write its complement 11.

(9 – 8 = 1; 10 – 9 = 1, infact 89 + 11 = 100).

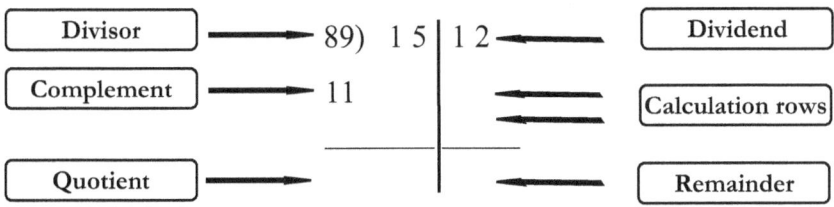

Return the first digit of the divisor (1) to the quotient line, then multiply this value by the complement (11).

$$1 \times 11 = 11$$

The product is reported in the second row, writing it down by scaling one column, starting below 5:

Note: *The remainder and intermediate products must have a number of digits equal to the zeros of the reference base (base 100: 2 digits, base 1000: 3 digits, etc.). For example, if a value of three figures is obtained instead of two, the first two digits on the left are grouped as tens.*

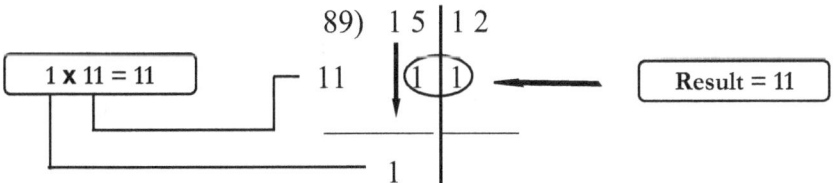

Summing in the column, 5 + 1 = 6, returning the sum to the quotient row, we get:

$$
\begin{array}{r|l}
89) \ 1\,5 & 1\,2 \\
11 \quad 1 & 1 \\
\hline
1\,6 &
\end{array}
$$

Multiply 6 **x** 11 = 66. Write the value in the next line of calculation by scaling one column:

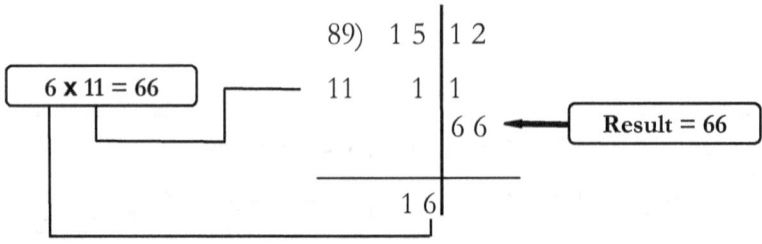

Summing in the columns: column of tens: $1 + 1 + 6 = 8$ and column of units: $2 + 6 = 8$, reporting the sums on the remainder row, we get:

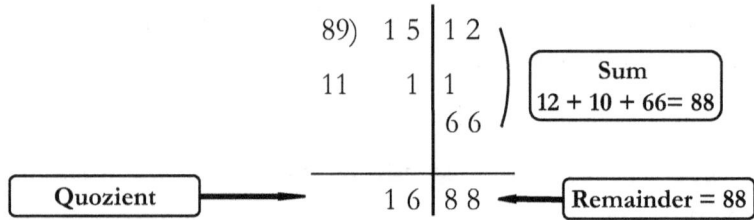

Result: 16   remainder: 88

Example:

$$1563 \div 88$$

88 is close to 100, and its complement is $100 - 88 = 12$.

Place 88 to the left of the pattern and 1563 to the right. Then, separate two figures from the right since the base is 100.

The remainder of the division will have at least two digits.

Under the divisor 88, write its complement 12.

$$
\begin{array}{r|l}
88) \ \ 1\,5 & 6\,3 \\
12 & \\
\hline
&
\end{array}
$$

Return the first digit of the divisor (1) to the quotient line, then multiply this value by the complement (12). The product is reported in the second row, writing it down by scaling one column, starting below 5:

$$88) \quad 1\ 5\ |6\ 3$$

| 1 x 12 = 12 | 12 | 1 | 2 | Result = 12 |

$$1$$

Summing in the column, $5 + 1 = 6$, returning the sum to the quotient row:

$$88) \quad 1\ 5\ |6\ 3$$

$$12 \quad\quad 1\ |2$$

$$1\ 6|$$

Multiply $6 \times 12 = 72$. Write the value in the next line of calculation by scaling one column:

$$88) \quad 1\ 5|\ 6\ 3$$

| 6 x 12 = 72 | 12 | 1 | 2 | |
| | | | 7 2 | Result = 72 |

$$1\ 6\ |15\ 5$$

After adding the following columns: $3 + 2 = 5$ and $6 + 2 + 7 = 15$, report the sums on the remainder row.

In this case, **the remainder is greater than the divisor.** Since 88 is contained once in 155, and the remainder is 67, add 1 to the quotient and leave the remainder equal to 67. (*Quickly, we subtract 88 from the remainder of 155 and add 1 to 16*).

88) 1 5 | 6 3

12   1 | 2
                | 7 2

_____

            1 6 | 15 5
            1 7 | 6 7

17 remainder = 67

> 155 greater than 88
> 155 − 88 = 67
> 16 + 1 = 17

Division above a base

Similarly, we can perform divisions with divisors near and **above a base**.

In this case, under the divisor, instead of the complement, we will write the difference with the base.

For example, if the divisor is 104, in the column below the divisor, we will write the excess 04, but with a **negative** sign.

(104 + (− 4) = 100).

The rule is: "Transpose and apply" because you transpose the excess from the base and apply the same method.

When we use **negative** values, the calculation is simple if you represent these with **bar numbers** (*vinculum*).

Example:

$$1563 \div 104$$

104) 1 5 | 6 3

$\overline{04}$

_____

We rewrite 1 in the quotient line and multiply it by 04. We return the product (which must have two digits because the base is 100) to the calculation row:

80

$$104) \ 1 \ 5 \ | \ 6 \ 3$$
$$\overline{04} \quad \ 0 \ | \ \overline{4}$$
$$\underline{}$$
$$1 \ |$$

Sum vertically and multiply by $\overline{04}$: ($5 + 0 = 5$ and $5 \times \overline{4} = \overline{20}$).

Write the product (20) in the calculation row by scaling one column:

$$104) \ 1 \ 5 \ | \ 6 \ 3$$
$$\overline{04} \quad \ 0 \ | \ \overline{4}$$
$$\overline{2} \ \overline{0}$$
$$\underline{}$$
$$1 \ 5 \ |$$

Sum vertically ($6 - 4 - 2 = 0$ and $3 - 0 = 3$) and obtain the remainder 03:

$$104) \ 1 \ 5 \ | \ 6 \ 3$$
$$\overline{04} \quad \ 0 \ | \ \overline{4}$$
$$\overline{2} \ \overline{0}$$
$$\underline{}$$
$$1 \ 5 \ | 0 \ 3$$

15   remainder = 3

Example:

$$13546 \div 106$$
$$106) \ 1 \ 3 \ 5 \ | \ 4 \ 6$$
$$\overline{06}$$
$$\underline{}$$
$$1 \ |$$

Write 1, multiply by $\overline{06}$, and return the product to the calculation line:

$$106)\ 1\ 3\ 5\ |4\ 6$$
$$\overline{06}\qquad 0\ \overline{6}$$

$$1\ 3$$

Multiply $1 \times \overline{06} = \overline{06}$, then sum vertically $3 + 0 = 3$.

$$106)\ 1\ 3\ 5\ |4\ 6$$
$$\overline{06}\qquad 0\ \overline{6}$$
$$\overline{1}\ |\overline{8}$$

$$1\ 3\ \overline{2}$$

Multiply $3 \times \overline{06} = \overline{18}$, then sum the column $5 - 6 - 1 = -2 = \overline{2}$.

$$106)\ 1\ 3\ 5\ |4\ 6$$
$$\overline{06}\qquad 0\ \overline{6}$$
$$\overline{1}\ |\overline{8}$$
$$1\ 2$$

$$1\ 3\ \overline{2}\ |\overline{3}$$

Multiply $\overline{2} \times \overline{06} = 12$, then sum $4 - 8 + 1 = -3 = \overline{3}$ and $6 + 2 = 8$.

$$106)\ 1\ 3\ 5\ |4\ 6$$
$$\overline{06}\qquad 0\ \overline{6}$$
$$\overline{1}\ |\overline{8}$$
$$1\ 2$$

$$1\ 3\ \overline{2}\ |\overline{3}\ 8$$

The remainder is $= \overline{3}8$

The quotient is $13\overline{2} = 130 - 2 = 128$.

The remainder becomes $\overline{3}8 = -30 + 8 = -22$

(128 with remainder – 22 means that 13546 is smaller by 22 compared to 28 X 106. Therefore, only 127 X 106 is contained in 13546 and the remainder (– 22) must be increased by 106).

Since the **remainder is negative**, one unit of the quotient is added *(which, for the remainder, worth is 106)* to the remainder so that it becomes positive. The quotient is then decreased by 1:

Remainder = – 22 + 106 = 84   e   Quotient = 128 – 1 = 127

127   remainder = 84

Note: *Subtract from the remainder a value equal to the divisor* **when the remainder exceeds the divisor.** *Consequently, add one to the quotient.*

When the **remainder is negative,** *subtract one from the quotient and add a value equal to the divisor to the remainder to make it positive.*

Performed examples:

a)

43615 ÷ 96

Base 100, remainder, and products: 2 digits

```
96)  4 3 6 |1 5
04      1 6
         1 |6
           |5 2
        ___|_____
         1 |
       4 4 3|12 7
```

(127 – 96 = 31) e (453 + 1 = 454)

4 5 4 | 3 1

454   remainder = 31

b)

$$251242 \div 989$$

Base 1000, remainder, and products: 3 digits

```
989)  2 5 1 | 2 4 2
011      0 2 | 2
           0 | 5 5
             | 0 3 3
      ───────┼───────
             |   1
      2 5 3 | 9 2 5
      253   | 1025
```

(1025 − 989 = 36) e (253 + 1 = 254)

254 remainder = 36

c)

$$3215 \div 997$$

Base 1000, remainder, and products: 3 digits

```
997)   3 | 2 1 5
003      | 0 0 9
      ───┼───────
         |   1
      3 | 2 1 4
      3 | 2 2 4
```

3 remainder = 224

d)

$$326427 \div 93$$

Base 100, remainder, and products: 2 digits

```
93) 3 2 6 4 | 27
7     2 1   |
        2 8 |
          6 | 3
            | 126
    ────────┼──────
        1   |   1
    3 4 9 8 | 173
    3 5 0 8 | 18 3
```

84

$$(183 - 93 = 90) \quad e \quad (3508 + 1 = 3509)$$

3509   remainder = 90

e)

$$721 \div 102$$

Base 100, remainder, and products: 2 digits

$$
\begin{array}{c|c}
102) & 7 & 2\ 1 \\
\overline{02} & & \overline{1}\ \overline{4} \\
\hline
& 7 & 1\ \overline{3} \\
& 7 & 0\ 7
\end{array}
$$

7   remainder = 7

f)

$$472615 \div 1009$$

Base 1000, remainder, and products: 3 digits

$$
\begin{array}{r|l}
1009) & 4\ 7\ 2 & 6\ 1\ 5 \\
\overline{009} & 0\ \overline{3} & \overline{6} \\
& 0 & \overline{6}\ \overline{3} \\
& & 0\ 0\ 9 \\
\hline
& & 1 \\
& 4\ 7\ \overline{1} & \overline{6}\ \overline{2}\ 4 \\
& 4\ 7\ \overline{1} & \overline{6}\ \overline{1}\ 4 \\
& 4\ 6\ 9 & -\ 610 + 4 \\
& 4\ 6\ 9 & -\ 606
\end{array}
$$

$$(1009 - 606 = 403) \quad e \quad (469 - 1 = 468)$$

468   remainder = 403

g)

$$63254 \div 1011$$

Base 1000, remainder, and products: 3 digits

$$
\begin{array}{r|l}
1011) & 6\,3\ \ 2\,5\,4 \\
\overline{011} & \quad 0\ \ \overline{6}\ \overline{6} \\
& \quad 0\ \overline{3}\,\overline{3} \\
\hline
6\,3 & \overline{4}\,\overline{4}\,1 \quad (-440 + 1) \\
6\,3 & -\ 4\,3\,9 \\
6\,2 & 10011 - 439 = 572
\end{array}
$$

62   remainder = 572

h)

$$3256 \div 1003$$

Base 1000, remainder, and products: 3 digits

$$
\begin{array}{r|l}
1003) & 3\ \ 2\,5\,6 \\
\overline{003} & \quad 0\,0\,\overline{9} \\
\hline
3 & 2\,5\,\overline{3} \\
3 & 2\,4\,7
\end{array}
$$

3   remainder = 247

Exercises:

6.2 Practice the following divisions:

a) $2122 \div 98$ $\qquad$ $315 \div 97$ $\qquad$ $34561 \div 96$

98) 2 1 | 2 2 $\qquad$ 97) 3 | 1 5 $\qquad$ 96) 3 4 5 | 6 1

b) $73252 \div 998$ $\qquad$ $83419 \div 994$ $\qquad$ $612712 \div 989$

998) 7 3 | 2 5 2 $\qquad$ 994) 8 3 | 4 1 9 $\qquad$ 989) 612 | 712

$$\ldots\ldots - 899 =$$

_____|___

c) $312 \div 101$ $\qquad$ $4225 \div 106$ $\qquad$ $37421 \div 1003$

101) 3 | 1 2 $\qquad$ 106) 4 2 | 2 5 $\qquad$ 1003) 3 7 | 4 2 1

d) $213243 \div 1002$ $\qquad$ $2272331 \div 1005$ $\qquad$ $71581 \div 1011$

1002) 213 | 2 4 3 $\qquad$ 1005) 2 2 7 | 2 3 1 $\qquad$ 1011) 71 | 5 8 1

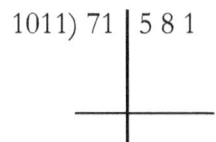

87

ANSWERS

1. Addition - Find zero and sum

Exercise 1.1

a) 25; 24; 21   *b)* 60; 91; 80   c) 90; 100; 2016; 1240

Exercise 1.2

a) 70; 55; 40; 71; 105   *b)* 80; 70; 200   c) 2; 2; 100; 1002,2

Exercise 1.3

15566; 21060; 22010

2. Subtractions

Exercise 2.1

a) 4; 73; 26; 57; 75   *b)* 762; 873; 210; 923
c) 7544; 2058; 9678;   *d)* 75,62; 673,16; 947,91

Exercise 2.2

a) 58; 141; 225; 190;   *b)* 543; 758; 1617; 1691

Exercise 2.3

a) 26; 16; 38; 27   *b)* 289; 478; 354   c) 1878; 2632; 1462

Exercise 2.4

a) 3649; 2224; 1466   *b)* 2782; 5294; 3386

3. Multiplying and dividing by 9, 11, 111

Exercise 3.1

a) 4122; 23202; 319239; 6541614
b) 49401; 689310; 8774667; 61078887

Exercise 3.2

a) 2695; 27896; 390071; 8103975
b) 72314; 909370; 10392635; 75628993
c) 69042; 91686; 83916; 76257

Exercise 3.3

   6 r 1;   13 r 5;   347;   472 r 5;   3151 r 4

Exercise 3.4

a)   7.9090...; 29; 89.7272...; 71.3636...
b)   240.4545....; 669.7272...; 899.6363...; 1320.2727...
c)   7128.3636....; 5157.6363...; 8434.4545...; 88033.3636...

Exercise 3.5

a) 8.7272...; 19.8181...; 71.4545...; 89.5454...
b) 225; 687,4545….; 803,3636...; 1415,7272...
c) 6229,4545...; 4704,9090…; 7516,2727...; 89759,7272...

4. Double and halve

Exercise 4.1

a) 86; 76; 158; 194   b) 482; 912; 1366   c) 6490; 4874; 15358

Exercise 4.2

a) 462; 658; 526; 1138   b) 6446; 8642; 4552; 13054

Exercise 4.3

a) 470; 648; 1464; 1908   b) 902; 1246; 1164
c) 7282; 25268; 126288

Exercise 4.4

a) 168; 212; 96; 344   b) 61; 82; 113.5; 71.5   c) 2568; 1752; 18; 162

Exercise 4.5

a) 21; 28; 37; 43.5   b) 143; 176.5; 235.5; 468
c) 663; 1658; 2573.5; 1136.5

Exercise 4.6

a) $5_16\ 8 \div 2 = 284;$ $7_13_16 \div 2 = 368;$ $8\ 1_18 \div 2 = 409;$
   $3_12\ 7_1 \div 2 = 163\ r\ 1 = 163.5$
b) $3_16\ 7_12 \div 2 = 1836;$ $4\ 7_12\ 6 \div 2 = 2363;$
   $5_11_13_17_1 \div 2 = 2568\ r\ 1 = 2568.5;$ $87_165_1 \div 2 = 4382\ r\ 1 = 4382.5$
c) $3_16\ 9_11_12 \div 2 = 18456;$ $8\ 9_15_17_16 \div 2 = 44788;$

$7_13_13_17_15_12 \div 2 = 366876 ;$ $8\ 5_14\ 7_17_11_1 \div 2 = 427385\ r\ 1 = 427385.5$

Exercise 4.7

a) $24 \times 5 = 12 \times 10 = 120;$ $32 \times 22 = 64 \times 11 = 704;$
   $45 \times 16 = 90 \times 8 = 720;$ $16 \times 14 = 32 \times 7 = 224;$
   $18 \times 35 = 9 \times 70 = 630$

b) $248 \times 2,5 = 124 \times 5 = 620;$ $125 \times 24 = 250 \times 12 = 500 \times 6 = 3000;$
   $48 \times 1,5 = 24 \times 3 = 72;$ $180 \times 3,5 = 90 \times 7 = 630$
   $64 \times 25 = 32\ \times 50 = 16\ \times 100 = 1600$

Exercise 4.8

a) $32 \times 5 = 320 \div 2 = 160;$ $17 \times 5 = 170 \div 2 = 85;$
   $422 \times 5 = 4220 \div 2 = 2110;$ $165 \times 5 = 1650\ \div 2 = 825$
   $318 \times 5 = 1590$
b) $28 \times 50 = 2800 \div 2 = 1400;$ 6250;   15600;   9100;   3200
c) $64 \times 15 = 640 + 320 = 960;$ 270;   3300;
   $24 \times 25 = 2400 \div 4 = 600;$ 1800
d) 3200;   8125; $412 \times 0.5 = 412 \div 2 = 206;$ 91;   128

Exercise 4.9

a) $42 \div 5 = 4.2 \times 2 = 8.4;$ $88 \div 5 = 8.8 \times 2 = 17.6;$ 64.2;
   $315 \div 25 = 3.15 \times 4 = 12.6;$ 24.72
b) 3.64;   5; $62 \div 50 = 124 \div 100 = 1.24;$ 2.84;   2.22
c) $32 \div 0.5 = 32 \times 2 = 64;$ 122;   448;   884;   198

5. Fast multiplications

Exercise 5.1

a) 2025; 1216; 7221; 9025   b) 216; 4224; 2009; 1221
c) 225; 2025; 4225; 5625; 13225   d) 11025; 11016; 15625; 15624

Exercise 5.2

a) 2925; 1909; 2964; 3481   *b)* 1701; 3036; 1316; 2409
c) 2581; 2501; 2709; 1209

Exercise 5.3

a) 2625; 2375; 1575; 875   *b)* 1425; 2125; 2275;
c) 2375; 2625; 4675;

Exercise 5.4

a) $97 \times 99 = 97 - 1 \mid \overline{3} \times \overline{1} = 9603$; $86 \times 98 = 86 - 2 \mid \overline{14} \times \overline{2} = 8428$

$91 \times 89 = 91 - 11 \mid \overline{9} \times \overline{11} = 8099$; $88 \times 95 = 88 - 5 \mid \overline{12} \times \overline{5} = 8360$

b) $92 \times 81 = 81 - 8 \mid \overline{8} \times 1\overline{9} = 73^{1}52 = 7452$;

$89 \times 75 = 75 - 11 \mid \overline{11} \times \overline{25} = 64^{2}75 = 6675$

$87 \times 97 = 87 - 3 \mid \overline{13} \times \overline{3} = 8439$;

$84 \times 88 = 84 - 12 \mid \overline{16} \times \overline{12} = 72^{1}92 = 7392$

c) $988 \times 994 = 988 - 6 \mid \overline{12} \times 00\overline{6} = 982072$;

$985 \times 996 = 985 - 4 \mid \overline{15} \times 00\overline{4} = 981060$;

$998 \times 875 = 875 - 2 \mid \overline{2} \times \overline{125} = 873250$;

$989 \times 888 = 888 - 11 \mid \overline{11} \times \overline{112} = 877^{1}232 = 878232$

Exercise 5.5

a) $102 \times 104 = 104 + 2 \mid 2 \times 4 = 10608$;

$111 \times 103 = 111 + 3 \mid 11 \times 3 = 11433$;

$1001 \times 1009 = 1009 + 1 \mid 001 \times 009 = 1010009$;

$120 \times 115 = 120 + 15 \mid 20 \times 15 = 135^{3}00 = 13800$

b) $111 \times 106 = 111 + 6 \mid 11 \times 6 = 11766$;

$1012 \times 1011 = 1012 + 011 \mid 012 \times 011 = 1023132$;

$113 \times 103 = 113 + 3 \mid 13 \times 03 = 11639$;

$1125 \times 1002 = 1125 + 002 \mid 125 \times 002 = 1127250$

c) $135 \times 111 = 135 + 11 \mid 35 \times 11 = 146^{3}85; = 14985;$

$123 \times 105 = 123 + 5 \mid 23 \times 05 = 128^{1}15 = 12915;$

$1250 \times 1011 = 1250 + 011 \mid 250 \times 011 = 1261^{2}750 = 1263750;$

$1012 \times 1106 = 1106 + 012 \mid 012 \times 106 = 1118^{1}272 = 1119272$

Exercise 5.6

a) $102 \times 99 = 102 - 01 \mid 2 \times 0\overline{1} = 1010\overline{2} = 10098$

$111 \times 89 = 111 - 11 \mid 11 \times \overline{11} = 100^{\overline{1}}\overline{21} = 99\overline{21} = 9879$

$112 \times 88 = 112 - 12 \mid 12 \times \overline{12} = 100^{\overline{1}}\overline{44} = 99\overline{44} = 9856$

$120 \times 75 = 120 - 25 \mid 20 \times \overline{25} = 95^{\overline{5}}\overline{00} = 90\overline{00} = 9000$

b) $113 \times 96 = 113 - 04 \mid 13 \times 0\overline{4} = 109\overline{52} = 10848$

$1012 \times 998 = 1012 - 002 \mid 12 \times \overline{002} = 1010\overline{024} = 1009976$

$1014 \times 998 = 1014 - 002 \mid 14 \times \overline{002} = 1012\overline{028} = 1011972$

$1025 \times 996 = 1025 - 004 \mid 25 \times \overline{004} = 1021\overline{100} = 1020900$

c) $105 \times 97 = 105 - 03 \mid 05 \times 0\overline{3} = 102\overline{15} = 10185$

$1115 \times 996 = 1115 - 004 \mid 115 \times \overline{004} = 1111\overline{460} = 1110540$

$1250 \times 998 = 1250 - 002 \mid 250 \times \overline{002} = 1248\overline{500} = 1247500$

$1012 \times 997 = 1012 - 003 \mid 012 \times \overline{003} = 1009\overline{036} = 1008964$

Exercise 5.7

a) $47^{2} = 25 + (-3) \mid 09 = 2209;$ $\qquad$ $54^{2} = 25 + 4 \mid 16 = 2916$

$39^{2} = 25 + (-11) \mid {}^{1}21 = 1521;$ $\qquad$ $62^{2} = 25 + 12 \mid {}^{1}44 = 3804$

$41^{2} = 25 + (-9) \mid 81 = 1681;$

b) $38^{2} = 25 + (-12) \mid {}^{1}44 = 1444;$ $\qquad$ $53^{2} = 25 + 3 \mid 09 = 2809$

$44^{2} = 25 + (-6) \mid 36 = 1936;$ $\qquad$ $56^{2} = 25 + 06 \mid 36 = 3136$

$43^{2} = 25 + (-7) \mid 49 = 1849$

Exercise 5.8

a) 121878;   212787;   231768;   741258

b) 511488;   956043;   103896;   624375

c) 404595;   11148885;   12498750;   10118988

Exercise 5.9

a) 8181;   2727;   3232;   7474

b) $52^{1}04 = 5304$;   4824;   7272;   4182

c) $412^{4}12 = 41612$; $315^{6}30 = 32130$; $250^{2}50 = 25250$;   1133

Exercise 5.10

a) 728;   715;   1225;   1221;   1232;   4209

b) 2125;   3136;   2736;   9212;   8099;   2419

c) 3330;   3995;   2058   1863;   3264;   6004

Exercise 5.11

a) 126846;   75114;   113515;   240192;   207662;   81496

b) 114911;   223033;   66543;   262485;   443360;   134875

c) 72360;   201656;   165312;   233358;   33072;   20925

Exercise 5.12

a) 49;   529;   2116;   3969;   7569;   4624

b) 15625;   54289;   116281;   196249;   394384;   984064

6. Fast divisions

Exercise 6.1

a) 10204; 30507; 3061   *b)* 7103; 90406; 10702

c) 105011; 201203; 20304   *d)* 11042

Exercise 6.2

a)

$2122 \div 98$	$315 \div 97$	$34561 \div 96$

```
98) 2 1 | 2 2        97) 3 | 1 5         96) 3 4 5 | 6 1
    2    0 4             3    0 9             4    1 2
         0 2            _____               2 0
    _____          3 | 2 4                    3 6
    2 1| 6 4                                  _____
                                              3 5 9 | 9 7  = 360 | 1
```

b)

$73252 \div 998$	$83419 \div 994$	$612712 \div 989$

```
998) 7 3 | 2 5 2      994) 8 3 | 4 1 9       989) 6 1 2 | 7 1 2
     2    0 1 4            6    0 4 8             11   0 6  6
          0 0 6                0 1 8                  0   1 1
     _____        _____                  0 8 8
     7 3 | 3 9 8          8 3 | 9 1 7          _____
                                               6 1 8 | 15 1 0
                                               619 | 1510 − 989 = 521
```

c)

$312 \div 101$	$4225 \div 106$	$37421 \div 1003$

```
101) 3 | 1 2         106) 4 2 | 2 5          1003) 3 7 | 4 2 1
 − 1     0 3̄          − 6    2̄ 4̄             − 3    0 0 9̄
 _____                   0 0                    0 2̄ 1̄
   3 | 1 1̄            _____           _____
   3 | 9              4 0 | 2̄ 5 = 40 | 1̄5     3 7 | 4̄ 9 0
                      39 | 106 − 15 = 91      37 | 310
```

d)

$213243 : 1002$	$227231 : 1005$	$71581 : 1011$

```
1002) 2 1 3 | 2 4 3       1005) 2 2 7 | 2 3 1        1011) 7 1 | 5 8 1
 − 2     0 0 4̄             − 5    0 1̄ 0              − 11    0 7̄ 7̄
         0 0 2̄                   0 1̄ 0                      0 1̄ 1̄
         0 0 6̄              _____             _____
 _____          2 2 6 | 1 0 1             7 1 | 2̄ 0 0
 2 1 3 | 2̄ 2̄ 3̄  = 213 | 1̄83                          70 | 1011 − 200 = 811
 212 | 1002 − 183 = 819
```

REFERENCES

Shri Bharati Krsna Tirthaji, *Vedic Mathematics*, published by Motilal Banarsidass, edn. 1992. ISBN 81-208-0163-6

Glover James, *The Curious Hats of Magical Maths: Vedic Magical Mathematics for school*, published by Motilal Banarsidass, edn. 1992. Book 1 ISBN 978-81-208-3973-1 - Book 2 ISBN 978-81-208-3974-8

Williams, K. R., *Vedic Mathematics Teacher's Manual Elementary Level*, Inspiration Books, edn. 2009. ISBN 978-1-902517-16-2

Williams, K. R., *Vedic Mathematics Teacher's Manual Intermediate Level*, Inspiration Books, edn. 2009. ISBN 978-1-902517-17-9

Williams, K. R., *Vedic Mathematics Teacher's Manual Advanced Level*, Inspiration Books, edn. 2009. ISBN 978-1-902517-18-6

Nolasco Pietro, *Il calcolo veloce*, Casa Editrice Kimerik, Ed. 2013, ISBN 978-88-6096-826-5

Nolasco Pietro, *Elementi di Matematica Vedica con esercizi: Corso base,* Pietro Nolasco, Ed. 2018, ISBN 978-8827572177

Nolasco Pietro, *Elementi di matematica vedica con esercizi corso intermedio,* Pietro Nolasco, Ed. 2020, ISBN 979-1220204996

Nolasco Pietro, *Esercitarsi con le moltiplicazioni. Impara velocemente con la Matematica Vedica: Esercizi di matematica per la scuola primaria, età 8 – 11 anni con 1502 quiz in 40 schede con soluzioni e valutazione,* Pietro Nolasco, Ed. 2021, ISBN 979-8534016802

Vedic Mathematics Academy, *www.vedicmaths.org*

Institute For The Advancement Of Vedic Mathematics *instavm.org*

Calcolo Veloce, *www.calcoloveloce.it*

If you liked the book, please leave a positive review.
If you have questions or want to learn more about the topic,
you can find material on
www.calcoloveloce.it